BUCKY'S DOME

BUCKY'S DOME

The Resurrection of R. Buckminster Fuller and
Anne Hewlett Fuller's Dome Home in Carbondale, Illinois

CARY O'DELL AND THAD HECKMAN

AMERICA
THROUGH
TIME

America Through Time
An imprint of Sutton Publishing Inc
www.through-time.com
office@through-time.com

First published 2020
Reprinted 2025

ISBN 978-1-63499-210-7

Typeset in 10pt on 13pt Sabon
Printed and bound in the United States of America

PRIMO:
A FOREWORD BY NORMAN FOSTER,
BARON FOSTER OF THAMES BANK

primo-: pri·mo (/ˈprē-(ˌ)mō/) n. the first or leading part.

A house, with 1,400 square feet of living space on two floors, was built from scratch using the strongest structural system known to man, climate-controlled for comfort at a time when air conditioning was uncommon. The cost was $60,000 at today's prices. Also, by the way, it takes twelve hours to build. This is not, of course, a futuristic vision of affordable housing, but a description of the house Buckminster Fuller—better known as "Bucky"—built for himself and his wife, Anne, on April 20, 1960. It was to remain his home while he taught at Southern Illinois University at Carbondale until 1971—the year I first met the man who continues to inspire and inform my own work to this day.

The Fuller Dome Home is unique among the surviving artifacts of Bucky's career. It was the first geodesic house he ever lived in, and the only home he ever owned. The eleven or so years he lived there coincided with a creative burst—in his writing, designing, and teaching—which was remarkable, even by Bucky's own extraordinary standards. Geodesics were refined and advanced in this house, and the wonders of tensegrity explored. All while, the students benefited from his semi-legendary extended lectures at SIUC.

One constant in the period, and throughout his career, was a concern for housing. At Carbondale, he devised plans for houses on artificial islands, houses under the sea, and even housing units that float. All these plans used aspects of geodesics, the system he originally worked out in the mid-1950s. Geodesics had superseded his "Dymaxion" house period of 1929–46, but in many ways, their concerns were the same: how to do more with less and how to mass manufacture cheap, strong, and lightweight homes.

I was fortunate enough to work with Bucky on a final iteration of the geodesic dome house. We planned to build two versions of the "Autonomous House" that we designed together—one for the Fullers in Los Angeles and one for my family in Wiltshire, England. Bucky's death in 1983 left the project unrealized, though the model we realized with the help of his assistant, John Warren, was displayed at my old studio in Great Portland Street, London, and is now housed in the Norman Foster Foundation in Madrid.

Lord Norman Foster with the 2011 Holland Prize Drawing: The R. Buckminster Fuller and Anne Hewlett Fuller Dome Home. (*Copyright Nigel Young / Foster + Partners*)

The design, like the Fuller Dome Home in Carbondale, was inspired by Bucky's life-long quest while making a minimal impact of the Earth and its resources. The Fuller Home Dome stands as a wonderful example of Bucky's philosophy of "doing more with less."

Bucky was forever warning about the fragility of the planet and our responsibility to protect it. For me, he embodied the very essence of a moral conscience. Though his public image was often that of a cold technocrat, in reality, nothing could be further from truth. The Bucky I knew was poetic, sentimental, and had a deeply spiritual dimension. He was one of those rare individuals who could fundamentally influence the way that you come to view the world.

I am delighted that the preservation of the entire structure is now nearly complete, and I pay tribute to the meticulous work of the team in restoring the home to its former glory. This was made possible by the creation of the not-for-profit Fuller Dome Home organization in 2002 by local volunteers. They succeeded in securing a Save America's Treasures Grant, subsequently matched in size by donations from many members of the public, to complete the project. I think Bucky would be touched by the commitment and generosity of everyone concerned. The restored Fuller Dome Home will play a crucial role in promoting Bucky's enduring message of hope for the future of humanity to a wider audience.

Lord Norman Foster is an acclaimed English architect, who founded the London-based architectural firm Foster + Partners. In his career, he has been awarded the Pritzker Architecture Prize, the Prince of Asturias Award, and the AIA gold medal for his contributions to the field.

ACKNOWLEDGMENTS

Writing this book has been a joy.

Our greatest gratitude to Bill Perk and Michael ("Mickey") Mitchell for preserving the Dome; and to Cornelius Crane, Brent Ritzel, and Jon Davey, the other great lynchpins in the saga of the Bucky house. This work, not to mention this home, could not have been completed without the assistance and candor of them all.

Also a big "Dymaxion" thanks to everyone interviewed for this work and who are quoted within it. Thank you for your ample assistance.

Our gratitude as well to the following: Allegra Fuller Snyder, the late Ben Gelman, Aaron Lisec, Tom Denton, John Lenzini, Robert Swenson, Matt Gorzalski, Gail White, Maureen Berkowitz, Hsaio-Yun Chu, Dede Ittner, Brad Dillard, the late Richard Archer, Jennifer Sorrell, Mary O'Hara, Bill Vollmer, Brenda Kirkpatrick, The SIU Alumni Association, The Stanford University Archives, The Buckminster Fuller Institute, and *The Southern.*

Thanks to all RBF Dome NFP board members, past and present.

Cary O'Dell would like to acknowledge the contributions of the following: Thad Heckman, Dolores Ford Mobley, Les O'Dell and family, Mike Heintz, and Karmon Runquist.

Thad Heckman would like to acknowledge the contributions of: Lord Foster, Cary O'Dell, Sierra, Asa, and Patty Heckman.

To others perhaps not listed but who also contributed—thank you.

Finally, of course, thank you to R. Buckminster Fuller and Anne Hewlett Fuller, without whom this book would not have been possible.

CONTENTS

INTRODUCTION

This book is a biography, but not of Buckminster Fuller. Instead, it is the story of a home—his home—specifically of its creation, inhabitation, endurance, dilapidation, and, finally, its restoration. It is often said, "If these walls could talk." Well, in this case, they do.

The "Bucky Dome" (as it is affectionately known among the locals) is the Carbondale-based one-time residence of Richard Buckminster Fuller and his wife, Anne Hewlett Fuller. The Fullers lived in the Dome during Bucky's ten-plus-year tenure as a professor at Southern Illinois University. It was recognized by the city of Carbondale as a historic landmark in 2003. It was added to the National Register of the Historic Places in 2006. Upon its final restoration, it is being fast-tracked to be named a National Landmark.

Yet as much as it is the story of the Bucky Dome (this "dwelling machine," as Bucky himself would have called it), it is also a story of people—the people who built it, lived in it, looked after it, saw to its safe-keeping, and then those who have worked to preserve it.

The group that has worked for over twenty years to rescue the Bucky Dome is the RBF Dome NFP. However, since its founding in 2002, the NFP has been concerned with far more than just nails, glass panes, and two-by-fours. The group's one-time president, Brent Ritzel, once said about their endeavor, "This project is about building community." For almost two decades now, the group has done just that. Along with rebuilding a house, the NFP also brought a lot of people together—people who knew or knew of Buckminster Fuller and had their lives forever touched by him and his passionate humanity. Each has their own story to tell, with the Bucky Dome serving as its common denominator.

R. Buckminster Fuller (1895–1983) was an architect, author, poet, engineer, designer, philosopher, educator, humanist, mathematician, futurist, and inventor. His structures cover more square footage on the planet than the work of any other architect or builder in recorded history. He was the recipient of America's highest civilian honor: the Presidential Medal of Freedom. He was nominated for the Nobel Peace Prize. Noted media theorist Marshall McLuhan once called him "the Leonard Da Vinci of our time."

During his time living in the Carbondale dwelling, Fuller would obtain eight patents, author ten books, and design his most famous work—a giant dome for Montreal's World Fair that served as the U.S.' pavilion. He would also teach a series of university-level classes and give countless lectures. Pertinent to his residency or not, during his decade at SIUC, the enrollment of the campus more than quadrupled—from 5,000 students when he joined to over 20,000 when he departed.

Obviously, the greatest significance of the Bucky Dome is that it was once the home of Buckminster Fuller and served as a fertile incubator for him during one of the most productive periods in his life. However, the home represents more than just its status as that. For seldom do the homes of the great and famous so vitally reflect the philosophy of their inhabitants, yet the "Bucky Dome" does.

First, the home is an *objet d'art*. It is as noteworthy on an artistic level as any other creation in the field of the visual arts during the highly inventive latter half of the twentieth century. Aesthetically, the home is as stunning today, in its retro futuristic way, as it was the day it was constructed. Now, recently restored with its striking original blue and white exterior, the home once again raises up unexpected and space-age, a bold pointer directing us towards a new architectural frontier.

Secondly, Carbondale's Bucky Dome also has the distinction of being one of the—if not the—very first residential domes. Today, over 200,000 dome homes dot the planet. Hence, Bucky and Anne's were among dome-living's very first test subjects.

Finally (and of course), the home epitomizes Bucky Fuller's greatest and most abiding legacy to the world: the geodesic dome. As the only dome home—in fact, the only home and property—that Bucky (and his wife, Anne) ever lived in and owned, the home is the most iconic and greatest surviving Fuller artifact. The home stands today as the embodiment of Bucky's philosophy, of the underlying concerns that drove all of his work. Rather than a message in a bottle, it is a message in a home. It is a message about eco-consciousness, of doing more with less, and of discovering long-lasting alternatives to tired traditions.

Symbolically, therefore, the home is also a testament to finding any sort of positive alternative to already-existing problems, and making sure that those solutions are systematically and democratically available to all.

In that regard, the Bucky Dome is a tribute and a communique to the future, as pertinent today as it was that day in 1960 when it first rose up from the ground on an empty corner lot at 407 South Forest Avenue in Carbondale, Illinois.

1

PROTO- (1960-1972)

proto-: pro·to (/ˈprōdō/) original; primitive; first; anterior; relating to a precursor.

Carbondale, Illinois, is located in the southwestern portion of Illinois. It is approximately 104 miles southeast from St. Louis and has a population of about 26,000. Since 1869, it has been the home of Southern Illinois University.[1]

Along with a multitude of accomplished graduates, over the decades, SIUC has also been the home to many highly accomplished and famous instructors. The great opera diva Marjorie Lawrence (whose life inspired the film *Interrupted Melody*) taught in the music department in the 1960s and '70s. In 1964, dance impresario Katherine Dunham became an instructor and "artist in residence" at the university; she stayed until 1967.[2] Later, Pulitzer Prize-winning novelist Richard Russo taught in the Department of English in the 1980s and early '90s.[3]

Arguably though, the school's most famous teaching alum is R. Buckminster Fuller, who was affiliated with SIU from 1959 until 1972.[4]

By the time he came to Carbondale, Buckminster Fuller—energetic, short, mostly bald, nearly blind, and sixty-four years old—was already being hailed as one of the great engineering and design visionaries of the century and was, along with other achievements, already the possessor of the longest entry in the *Who's Who* annual.[5] Mature in years and his thinking by now, Fuller had also by this time formulized some of his greatest breakthroughs.

In 1929, Fuller debuted his revolutionary hexagonal home—the 4D House. In 1933, he introduced his Dymaxion car—a three-wheeled, aerodynamic and fuel-efficient vehicle. In 1940, inspired in part by mass-produced metal grain silos, Fuller formulated Dymaxion deployment units (DDU). These units—low-cost, mass producible, and easy to move—were deployed in the hundreds by the military during the war years. In 1943, Fuller unveiled the design of his remarkable Dymaxion map, a flat representation of the Earth that, unlike most other cartological efforts, resulted in minimal geographical distortion. After the war, Fuller saw to the development of the Wichita Dwelling Machine.

Gaining its name from its initial location in Wichita, Kansas, this circular metal home was supported (like the earlier 4D home) by a central mast and was originally produced by Beech Aircraft using aircraft production techniques their factories had employed during the war.[6] The original Wichita House is still standing; it has, however, since been moved to The Henry Ford Museum in Dearborn, Michigan.[7]

However, even amid these great accomplishments, Buckminster ("Bucky" to his friends) Fuller was, by this time (and forevermore), most closely associated with the breakthrough known as the geodesic dome. It was in 1947 and 1948 that Buckminster Fuller, after hundreds of hours of tinkering, refined the design of the geodesic. Inspired by nature, and troubled by the U.S.' post-war housing shortage, Bucky began to ponder alternate shapes for architecture.[8] Robert R. Potter explained it in his 1990 biography of Fuller:

> [W]asn't the dome—a part of a sphere, after all—the lightest, strongest, and most logical structure a person could build? His map work had taught him about great circles. A great circle is the largest circle one can draw on the surface of a sphere.... Fuller discovered a way to put thirty-one great circles on a sphere so that the entire surface became a network of triangles. Cut such a sphere in half, you have two domes....[9]

Of all solid shapes, the sphere encloses the most space with the least amount of surface. Of all flat shapes, the triangle is the most rigid. So a dome made of triangles was both the most economical and the strongest of all building systems.

In practice, a geodesic dome, as a livable, useable structure, is created via the use of various short, straight struts, or chords, which, when fastened together at their ends along geodesic lines, creates a series of isosceles triangles—a triangle being geometry's most rigid shape—that, when then combined in a specific, symmetrically radial manner, form an open lattice work that is surprisingly sturdy though remarkably lightweight. The lattice, or framework, when continued and expanded, will create a curved structure—a dome—or, in the extreme, a full sphere.[10]

The use of spherical and half-spherical shapes had been part of buildings for millennia, at least since the time of ancient Rome. They were an outgrowth of arches and archways. However, early domes were heavy, and their weight had to be borne by means of reinforced support walls, flying buttresses, or other elaborate brick-and-mortar underpinnings. They were cumbersome at best, inefficient at worst. In contrast, the geodesic dome's mass and weight—no matter how big the dome gets—is spread out continuously across its frame via its own chords, its own self-supporting structure. A dome's own tension and compression forces, all balanced within its own geometric form, thus lends itself an innate resiliency and, as noted, a stunning lack of heft.[11]

Additionally, geodesic domes are incredibly quick to construct and relatively inexpensive to build. To Buckminster Fuller, they represented the full manifestation of his personal-professional creed: "Do more with less."[12]

Once Fuller conceived of, designed, and patented the mechanics of the geodesic dome—patent no. 2,682,235, issued in 1954—the economy, mobility, durability, and simplicity of the geodome was soon embraced by a wide fraction of groups.[13] Domes and dome-like structures began sprouting up all around the country and all around the

Right: R. Buckminster Fuller. (*Courtesy: SIUC/Morris Library*)

Below: Fuller with a cardboard model of a geodesic half-sphere.

world, performing a variety of uses. The U.S. military quickly utilized them as either temporary aircraft hangers or as short-term military housing. In 1956, the Canadian and U.S. governments utilized various domes—which they called "Radomes"—as part of their Distant Early Warning Line (or DEW Line) located near the Arctic Circle.[14]

Non-government entities were also fast to embrace domes. In 1953, the Ford Motor Company tapped Fuller to construct a dome as the new roof to their Dearborn, Michigan, rotunda. That same year, a dome restaurant opened in Woods Hole, Massachusetts. In 1957, Kaiser Aluminum and Chemical Corporation built two massive domes—one in Virginia and one in Hawaii. In 1958, the Union Tank Car Company put up a dome—which spanned more than a football field in area—near Baton Rouge, Louisiana.[15]

Geodesic domes have been called the most significant structural innovation in centuries. Today, they can be found on every continent (including Antarctica), as well as in almost every town. They have been adopted as everything from sports stadiums to band pavilions, from children's playground equipment to pop-up tents, and free-standing structures seen annually at Nevada's Burning Man Festival. Perhaps most famously, EPCOT Center's iconic central sphere (located near Orlando, Florida) is a geodesic dome design. In terms of residential construction, it is estimated that over 300,000 dome homes are currently providing living quarters for people all around the world.[16] According to the late Jay Baldwin, author of the book *BuckyWorks*, when tabulated, the sum of Fuller's domes collectively cover more of the Earth than the work of any other architect.[17]

As the dome's patent holder, Fuller partnered with various companies to bring domes to the masses. For domes to be used as homes, however, Bucky contracted with the Hamilton, Ohio-based Pease Woodworking Company. Together, Bucky and Pease began selling do-it-yourself dome kits. Sales were made through the company's literature and the company's team of roving salesmen. At Pease, to assist their fleet of reps, the company equipped each of their agents with a miniature dome model that they carried around with them inside their oversized *Mad Men*-era briefcases.[18]

Similar but more efficient than the famous Sears Roebuck homes that were available for purchase between 1908 and 1940 through the Sears catalog, Pease dome kits came with everything one needed (minus any permits and labor) to realize their own dome home. During their dome-selling heyday, Pease trumpeted the science and the many advantages of dome living and usage:

> Employing the geodesic principle developed by R. Buckminster Fuller, PEASE DOMES are based on mathematically precise division of the sphere—the strongest, most efficient system of structuring today. The PEASE DOME is an engineered system of triangular space frames, designed to distribute stresses equally throughout the building, the weight being transferred directly to the ground, instead of to land bearing walls, beams or partitions.[19]

Along with the Pease's A-7 model—the "3-room partition package"—which would serve as the model of the Fuller Carbondale home, the company also offered the A-5 style, their "one-room Vacation Dome"; the A-4, the company's "Holiday Dome"; the A-3 "Camper Dome" ("the new way of living outdoors"); the "Utility Dome" (for storage) and the "Canopy Dome" (a shady awning for picnics, etc.).[20]

An U.S. "Radome" built in the geodesic design.

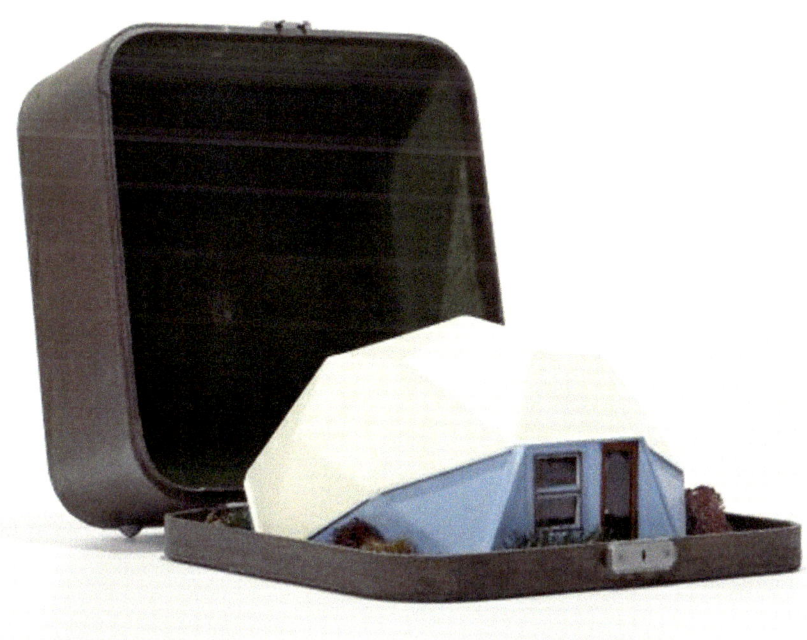

Sample case carried by Pease salesmen containing a model of one of their dome homes.

Along with a page of material specifications, the Pease brochure also touted the structure's "superior strength," as well as the dome's adaptability and affordability. The company's prices ranged from $390 for the Canopy dome to $1,450 for the A-7 model.[21]

Obviously, as the geodesic dome's creator and greatest promoter, it made thorough sense (or "synergy," as Bucky would have said) that Buckminster Fuller and his wife, Anne, would eventually build one. If domes were a good idea, then Bucky had to lead by example; if domes were the homes of the future, then the Fullers had to be on the vanguard. Additionally, for Pease, Bucky actually living in a dome would prove to be great advertising for them and their priced-to-sell, do-it-yourself dome packages. For Bucky, advocating the use of domes personally would help popularize their use for a reason he completely believed in—for the better sheltering of humanity.

Before coming to Carbondale, Fuller and Anne made their home on Long Island, in an apartment in Forest Hills.[22] Actually, more accurately, Anne Fuller made her home there as Bucky was often traveling to various teaching positions around the country or giving wide-ranging lectures at various far-flung locations all over the nation. For a time, for his lecturing, Bucky even crisscrossed the country in a car pulling a trailer in which he carried all his geodesic models.[23]

The Fuller family (Bucky, Anne, and their daughter, Allegra) resided in Forest Hills for fifteen years before being burned out of the location, an incident that thankfully did not harm anyone or destroy any of Bucky's documents.[24] By this time, Fuller's collection of writings, research notes, sketches, and drawings (which he labeled his "Chronofile") was mammoth. Finding appropriate, safe storage for it was one of the selling points that solidified Fuller's decision to accept a job offer as a research professor at Southern Illinois University in Carbondale.[25]

Bucky's arrival at SIUC and Carbondale was largely the work of two men: SIUC Design professor Harold Cohen and then SIUC President Delyte Morris. Cohen (a design notable in his own right) was the founding director of the school's Department of Design; he arrived in 1955. He had first encountered Fuller in 1948 when he attended a lecture of Bucky's held in Chicago. The two men became fast friends.[26] In 1956, shortly after the Forest Hills fire, Cohen received a desperate call from his friend, Bucky. Fuller was devastated by the fire and short on cash. Cohen came up with a solution—come and teach at SIUC.[27] After selling Fuller on the idea of becoming an academic, Cohen met with the university's president, Delyte Morris.

To "get" Bucky fully, Morris offered Fuller a one-of-a-kind position and an ample number of perks. Among them, Fuller would bear the title "World Fellow in Residence" and be allowed to teach in any department on any subject. He would also report only to Morris himself, making Fuller one of only four professors in SIUC history to be granted that status.[28]

Hence, in 1956, Buckminster Fuller accepted his first (and penultimate) academic appointment.

A few years later, in 1959, the same year that the university bestowed an honorary degree on him, Fuller was named a full professor by the university. Bucky's hiring announcement appeared in the press in September. In the article on his appointment in the *Southern Illinoisan*, it was noted that Fuller would begin work on October 5, by giving his first lecture to design students.[29] In his annual fall letter to students' parents

Above left: Cover of one of the brochures from the Pease Company promoting their line of domes.

Above right: Delyte Morris, the SIUC President who was instrumental in bringing him to Carbondale, poses alongside Fuller. (*Courtesy: SIUC/Morris Library*)

that year, Delyte Morris touted the addition of Fuller to the university's teaching staff as one of the major accomplishments for the school's upcoming term.[30]

Once hired by SIUC (at a salary of $12,000 a year), Fuller settled in to lecturing in the university's Department of Design (now part of the Department of Art and Design). In addition to his teaching salary, SIUC would also provide Fuller with his own office, staff, and workspace to conduct his experiments. Bucky would also be given a liberal travel schedule which allowed him to continue his active, global speaking schedule. In fact, Fuller's agreement with SIUC would only require him to be physically on campus for two (non-consecutive) months out of each year. At those times, Fuller would present lectures and conduct various student seminars in design. The rest of the time, Fuller was free to do, and come and go, as he wished.[31]

He did just that. All during Fuller's SIU years, Bucky kept up an extraordinary travel schedule. During 1959 and 1960, Fuller traveled to give speeches (often for as much as $1,000 per lecture) to such distant places as Massachusetts, Texas, California, and New Mexico.[32]

During his lifetime, it was estimated that Fuller traveled over 3 million miles. Some of the places Fuller would go to during the 1960s included Edinburgh, Brooklyn, Miami, Austria, Kansas, Australia, South Dakota, Chicago, Colorado, Kansas, Iowa, and Amherst.[33] Twice during his SIUC tenure, Fuller also held teaching positions with other colleges—one in Ghana and one at Briston University in the U.K.[34]

In 1967, the *Southern Illinoisan*'s Ben Gelman (long on the paper's "Bucky Beat" and, in time, a great friend of the Fullers) recounted Fuller's travel schedule for just May through June 1967: Fuller began, May 1, in Canada; on May 2, he was to fly to New York; on the 4th, he would be in Beirut, Lebanon; on the 7th, he would be in Bagdad; on the 11th, he would be in Syria but then head back to London on May 14; on the 17th, he was due in Arizona before going onto Salt Lake City and before then heading back to New York. On the 22nd, Fuller would be in Carbondale to speak. Fuller would remain in Carbondale until the 25th when he would then fly to Minneapolis to discuss plans for an experimental city (i.e. his East St. Louis "Old Man River City" project). Fuller would be back in Carbondale on the 28th before heading to New York for a TV appearance on the 30th. Fuller would then be in Maine for the first part of June but in Edwardsville, Illinois, by June 7. He would then be in Carbondale from the 8th to the 10th before heading out to the East Coast for an engagement at Harvard.[35]

Often accompanying Bucky was his dedicated Carbondale-based secretary, the late Naomi Smith (later Wallace). Smith had the joyful but daunting task of constantly taking down Fuller's dictation and then transcribing it. The results usually became the first draft of one of his books. Smith once said about her "on-the-go" career, "I kept a packed suitcase by my door. I never knew when I'd get a call to get on a plane and be in New York. I was on call constantly—and I loved it!"[36]

Yet amid this hectic itinerary, Fuller continued to produce and innovate. He did this, in part, thanks to a dedicated local staff. The size of Fuller's SIU staff would fluctuate widely over the years; it ranged from as few as nine to up to twenty at some points. Some employees of the office were paid directly by the university; others were paid by Bucky himself, though many of those funds came by way of an SIUC stipend given to Fuller. Some of his most important employees included Smith, Dale Klaus, Tom Turner, Joan Eubanks, John McHale, Evelyn Benson, and Diana Wachter, among others.[37]

Fuller's first office in Carbondale was located at 711 ½ South University, off campus, on the second story of a building that housed a bookstore and a travel agency. Later he would move to 206 West College Street.[38]

In either location, Fuller's office was a beehive of activities. It would be even further abuzz when Bucky himself was there. In his book, *Cosmic Fishing*, author E. J. Applewhite reported a typical day at Fuller's office:

> His office door was always open and there was always some delegation or another waiting for their turn in his presence. Michael was in charge of the research files and craved 15 minutes to brief Bucky on some exciting new development in physics and chemistry. Tom was waiting to get approval of a new agreement on the SIU Library custody of Fuller's voluminous archives. Shoji Sadoe was … calling that morning with questions about a Fuller dome being erected in Edwardsville. Dale needs guidance on map sales and new reprints of Fuller speeches. Don wants Bucky to address a national conference on nutrition in Washington next month. Herbert calls….[39]

Often, there were as many as a dozen different projects going on at any particular time. For example, the year 1960 also saw Fuller propose two of his most audacious engineering projects. First, Bucky suggested building a dome over the city of Manhattan. This clear half-sphere would have covered New York island from the East River to the Hudson, from 21st Street to 64th Street—a span of about 2 miles. The suggested dome would regulate the weather and reduce air pollution. Though, obviously, never constructed, such an epic dome is completely feasible.[40]

Also in 1960, Fuller put forth his idea of "Floating Cloud Structures." These spheres were another attempt by Fuller to find new living spaces for the ever-expanding populace. According to Fuller, these in-the-sky globes would be floated above *terra firma* by as much as 1 mile as long as the interior temperature of them was constantly kept at least 1 degree warmer than the air outside of it.[41]

No matter where he was or what else he was doing, Fuller was always writing (in longhand) or dictating new works. During his SIUC tenure, Bucky would write and publish ten books. His *The Dymaxion World of Buckminster Fuller* began the fusillade, appearing in 1960. This would be followed by *Tensegrity* in 1961. Remarkably, seven titles by Fuller appeared in 1963; they included *No More Second Hand God*; *Ideas and Integrities*; *Energetic-Synergetic Geometry*; *Education Automation*, and a reissue of his earlier *Nine Chains to the Moon*. Later, there would be *Utopia or Oblivion* in 1969; *I Seem to Be a Verb* and *Intuition* in 1970; and, finally, *Buckminster Fuller to Children of Earth* in 1972.[42]

Despite Bucky's infrequency on campus, he gave much to his students during the times he was in Carbondale. Rare was a Fuller lecture that lasted less than four hours, and though many took place in packed-to-capacity auditoriums, some were just as likely to be held "out in the field."

Fuller also found time during his short stays in Carbondale to give back to the campus and the community. For example, in 1962, Fuller was a judge on a scenic design contest put on by SIUC's Theater department.[43] In 1963, Bucky gave a free lecture in Browne Auditorium.[44] In May 1964, he spoke to a general audience of over 400 for over three hours on the topic of "World Design."[45] In November 1964, he spoke to over 300 art educators at SIUC's Student Center.[46] In 1965, Fuller gave a free lecture on "The Prospect of Humanity" in the SIUC's Wham Building and also addressed a local meeting

Fuller with students at SIUC.

of the Midwest Regional Conference of the American Business Writing Association.[47, 48] In October 1965, he was part of SIUC's multi-day "Vision '65" which welcomed over 200 guests to address "International Challenges for Human Communications."[49]

In 1967, Fuller spoke on the topic of "Grand Strategies of Realistic Thinking" to the National University Extension Association in SIUC's River Room, and one evening, he spoke to the local Carbondale branch of the PTA.[50, 51]

In 1971, Fuller took part in SIUC's ambitious "Alternatives '71" event. This sixteen-day gathering was staged to "emphasize the university and the community as a cultural entity." Part of "Alternatives" featured Bucky at an outdoor speaking event titled "Picnic with Bucky" that was held on the lawn by the stadium on SIUC campus.[52]

When Bucky first moved to Carbondale, he lived with Harold Cohen and Cohen's family. For the Cohens' children, Bucky used to tell them a bedtime story he titled "Gnomes That Live in the Dome." Anne Fuller visited often but would not move permanently to the area until the couple had their own house to live in.[53] Soon, the Fullers would have a dome all their own.

Only one month after his appointment to SIUC, Fuller stated in the press that "This will be my home base. I'm putting down roots in Southern Illinois." Part of those roots were going to be his own geodesic dome. Fuller said, "I have not sharpened up any plans for the house yet, but I will probably use the 55-foot Kaiser dome."[54]

For their new home, SIU President Delyte Morris directed the Fullers to a parcel of land on the corner of the city's South Forest and West Cherry Streets. The lot (lot no. 78 in University Place Addition; plot number 15-21-158-016-0040; 407 South Forest Street) measures roughly 51 by 121 feet. At that time, it was owned by Ralph and Anna Gray of West Frankfort and was vacant except for a pink mobile home parked there and rented out.[55]

Though near the center of the city, this small plot of land is nevertheless very much located in a residential neighborhood with the majority of the other homes on the block—all one- and two-story affairs—having been built in the 1930s and '40s. Today, the neighborhood, known as the Arbor District, is an array of student rentals and long-time Carbondale-ites. In Bucky's day, the area housed a mix of professors and other university staff. Even before the Fullers arrived, however, that part of town had an interesting pedigree. The birth home of notable newspaperwoman and long-time SIUC benefactor Virginia "The Duchess" Marmaduke stood (and still stands) only three blocks away.[56]

Before ground could be broken for building the Bucky Dome, the Dome had to be located on the site, its shape laid out within the lot, and a solid perimeter foundation wall had to be constructed. So, eight days before the building of the actual Dome shell, Carbondale's Parrish Construction came and poured a concrete foundation wall with a 4-inch thick slab and perimeter concrete foundation system. Its final shape was a ten-sided polygon that, when finished, actually appeared rather circular. It was on this concrete pad that the Dome would eventually be erected.[57]

Also poured that day was the driveway. The driveway sits lower than the home itself which then required Parrish to pour a retaining wall about 30 inches tall and then craft three steps that led upward to the house.[58]

The dome-building package from Pease provided all that was needed for this dome to be fully assembled. This included prefabricated wall and roof sections, a set of more conventional-looking doors and windows, and even some kitchen and bathroom

As the foundation for the Dome is poured, a worker surveys the building plans. (*Courtesy: SIUC/Morris Library*)

fixtures. The building's main sections arrived via sixty prefabricated isosceles triangles. The sixty triangular frames were comprised of six pentagons and five hexagons. The six pentagons (or "pents") are composed of five triangles that measure (approximately) 6 feet 9 inches at the matching edges with the base edges measuring 7 feet 10 inches. The building's hexagons are composed of six isosceles triangles of which the matching legs are close to eight feet with a base length just a fraction over 7 feet 10 inches. Each section consisted of a 3/8-inch-thick piece of unpainted plywood over a wooden frame made of two-by-three lumber cut precisely to the unique angles of a geodesic dome. Once unpacked, these pieces would then be aligned and bolted together on-site to form a dome that measured a little over 39 feet across and 16.5 feet tall at its highest point (the so-called "polar-pent").[59]

The finished dome would be a modest one-bedroom home of roughly 1,100 square feet on the ground floor. For Bucky and Anne's home, an additional 400 square feet (roughly) of additional living space was immediately created with the construction of an internal platform that, when installed, would serve as the home's second story. This upper space or "loft" (which would cover just under half of the Dome's main floor space) would be accessible by a rather narrow stairway. It would serve as Bucky's home study or, in a pinch, could be used as a second bedroom.[60]

A radiant hot water circulating heat system, achieved through a matrix of copper pipes, was embedded in the concrete slab and ran throughout the entire area of the home. This concept was one pioneered or at least advocated by Fuller. Due to this efficient method of heating, in the beginning of the Dome's lifespan, the doors could actually be left open during the winter months and the temperature inside would remain toasty warm.[61]

Left: Workers aligning one of the first triangular frames with the home's foundation.

Below: A workman installs the radiant heating system in the Dome's floor. Note the presence of the metal Pease Dome plate at the top of the stud near the top of the photo.

Air conditioning was not installed. However, air circulation and temperature uniformity were achieved via natural airflow patterns that were heightened within the Dome's structure. The upstairs was cooled and heated through pegboard/ventilation holes incorporated into the wooden baseboard of the loft's semicircular, room-length bookshelf. Below that were ceiling grilles in the kitchen and north bedroom that enabled air to circulate through the Dome. An operational awning window was placed on the (first floor) north wall of the bedroom, directly below the loft circulation holes, further enabling air movement. Finally, continuous whole house circulation was achieved thanks to a central electrical fan suspended from the center polar-pent at the top of the Dome; it would pull air through the ceiling grills and ventilation holes and circulate the air through in a giant "loop" fashion. Not to be confused with the ceiling fans of today, this fan was an industrial model, which was a little over 1 foot wide and encased in a wire cage. It was suspended from the center of the Dome by a metal, U-shaped harness and could be tilted to direct air flow at a desired angle. The on/off switch for the fan was on the wall just left of the bedroom window while the home's original thermostat was positioned in the living room.[62] The thermostat was, at some point, repositioned to a wall in the entryway where it has remained.[63]

Though the Fullers seemed to be quite comfortable in their home during their time living there, later residents often had trouble heating and cooling the home. By today's standards, the Dome was not well insulated when it was built. Originally, it had only a 1–2-inch layer of lightweight batt insulation glued with black mastic to the inside of each of its triangular panels.[64]

As for light, up near the central apex of the ceiling, ten overhead skylights, set flush with the roof surface and made of translucent white fiberglass, illuminated both the study loft and the main floor of the house.[65]

Also included in the Pease package were several large wheels of plastic tape to be used as a seal over all of the Dome's rooftop joints. The tape was made of Celastic soaked in methyl-ethyl-ketone (MEK) to make it pliable—able to be contoured to the unusual shapes of a geodesic dome. According to the manufacturer, once this tape was put in place, it would form a "rock hardness" and be waterproof.[66]

Along with being (allegedly) watertight, the house was, according to Pease, also miraculously strong. Company tests showed that the top of the Dome could withstand up to 26,500 pounds of direct pressure.[67]

Though Bucky and his wife might have gotten a deal, for other customers, this dome package would cost about $3,700. Combined with the cost of the foundation and water and electrical installation, the total for the home was probably between $7,000 and $8,000 (or about $1.25 per square foot) in 1959 dollars.[68] That would be about $60,000 today.

While, for most homes, construction can take weeks or months, the Fuller home's dome shell was erected in just one day—April 20, 1960.[69] The home's components were trucked in by Pease that morning. In a widely reproduced photo from that day, Bucky can be seen already on site, chatting with the company's truck driver.

Once again, Ira Parrish of Parrish Construction performed the Dome's erection. The late Mr. Parrish had been in business in the area since 1956. By the time they were tapped to assemble the Dome, Parrish and his firm had already built hundreds of homes as well as the original grandstand on the DuQuoin fairgrounds. The late Mr. Parrish

Above: Workmen apply Celastic tape to the seam between two of the home's wooden triangles.

Below: Fuller (right) chats with the truck driver who brought his home, in pieces, to the construction site the morning of the build.

recalled, "I knew [SIU President] Delyte Morris. I was in his office one day and he, more or less, hired me to put up the Dome."[70]

Neither Mr. Parrish nor any of his five-man crew (Walter Shewmaker, Kenneth Simpkins, Milton Harkins, Everet Lindsey, and Paul Parrish) had ever put up a dome before, but nevertheless, they felt confident in what they were taking on. The Pease package came with instructions and Bucky himself was going to be on-site. After the Pease pieces arrived, Parrish and his men pulled them from the truck and, following the instructions and the numbers imprinted on each of the pieces, bolted them together. As the Dome reached upwards, the workers employed a temporary scaffolding as well as several short-term wood posts to aid their work. The home formed its shape with surprising ease and speed.[71]

After the exterior shell was completed and the inside walls were framed, the Dome itself was lined with thin plywood panels cut to match the triangles. Then, each edge of the plywood was covered with half-round pine trim and painted. The interior walls were fitted with gypsum board, or "Sheetrock," and all the walls and plywood liner panels and wood trim was painted white. The final flooring finish installed over the concrete slab was cork.[72]

Cork (a mid-century design staple) was installed in pre-cut 12 × 12-inch tiles on the main level and in 12 × 6-inch pre-cut tiles up in the loft. On both levels, the cork would provide both cushion and insulation. These tiles were purchased from another local vendor. A painted wood base trim hid the joints between the floor and the Dome was installed after the floor's installation. On the first floor, the cork's thickness was about half an inch and colored a dark brown. The cork up in the loft was slightly thinner, about 1/8 of an inch, and was left untreated thereby retaining its original sandy-brown shade.[73]

According to Mr. Parrish, the only surprise of the day was the building and incorporation of the "box-like" interior structure that created the house's second floor loft and, when installed, would also hide away some of the home's mechanical systems. The additional wood for the second story was purchased from Associated Lumber, one of Carbondale's two lumberyards at the time.[74]

Work began on the Dome's shell just before 10 a.m.; it would be concluded by 5 p.m.[75] As the day ended and some builders began work on the interior, some of Parrish's other workers created the small water feature that sits at the south portion of the lot, about 30 feet from the south edge of the house. Fuller designed the front yard fountain, and it adds yet another futuristic feel to the home. Measuring 10 feet across, the water feature is a shallow, in-ground circle with a single spout in its center that shoots upwards a continuous spray. Originally, it is believed, Bucky had the fountain painted blue to match the color of the house.[76]

Parrish also began work on the large redwood fence that would eventually surround the property and close off the driveway from the street. The finished fence, comprised of (mainly) 8-foot-long sections, would stand a full 6 feet tall—and sometimes higher depending on grade.[77]

The finished fence creates a rectangular box around the property. It is completely square with right-angle corners, except for the southwest corner where the fence panels of the south and west are at a 45-degree angle; this allows vehicles coming around the corner to have a complete view of the intersection.[78]

Above: As the home rose from the ground during the day, a temporary scaffolding had to be erected by workers.

Left: The Dome takes shape.

Though the finished fence encloses the house, yard, and driveway, a shortened piece of fence was also put up as a divider of sorts—a separator between the driveway's entrance and the north-facing side of the home.[79] Added a little later, nestled into that northwest corner, was a small triangular shed, created for the storage of a lawnmower.[80]

Bucky designed the fence himself. It consisted of a series of closely arranged "V" slats that were angled in one direction for one half of the post-to-post span, then were angled in the reverse direction for remaining other half (and so on and so on). This design allowed for air to freely pass through the fence regardless of the wind's direction. It also blocked views into the yard.[81] Something of an engineering marvel, along with granting the Fullers a constant breeze and privacy, the fence was also a legal necessity—to have the small fountain out front, a local ordinance required it to be properly enclosed.[82]

Though Parrish and his men started the fence that day, they had to return later to finish it; not enough redwood was on hand in Carbondale and more had to be shipped in.[83] The cost of this made-to-order fence set the Fullers back about $2,500.[84]

In front of the Fullers' large driveway (a driveway large enough to accommodate up to four vehicles at one time), there was also erected a gate. Reaching twenty feet across, the gate was actually a series of four gates. Two gates spanning half the driveway operated separately on pivots that, when opened together allowed complete access to the drive. However, two "man-size" sub-gates with separate "cane bolts" were literally embedded within the large gates and fold back on the large gate. A clever design fitting Bucky, these sub-gates allowed pedestrian access to and from the street without the need for opening the entire large gate assembly.

Bucky, wearing a favorite hat, poses with his quickly evolving home.

Also constructed that day was a second, much smaller gate. Situated south of the home's front door, the gate was affixed to the house on one side and another part of the fence on the other. Waist-high, picketed, and made of the same redwood, this second gate, when closed, largely sealed off the yard area.[86]

Finally, two bollard lights were also installed. One is at each end of the wide steps that lead from the driveway into the home. They light the way after dark without glaring into the eyes. Each post is about 1.5 feet tall and each is topped with a grey, metal, half-dome shade (about 8 inches tall and 16 inches around) that looks a tad like a coolie hat.[87]

Though some years later, Parrish and crew would build a small canopy dome for the city of Carbondale, Fuller's residence is the only full dome Parrish and company would ever construct. Parrish said later, with a chuckle, "If I knew what a piece of history it was going to be, and how much I would be asked about it, I would have paid more attention that day."[88]

Throughout that eight-hour day of construction, Bucky was on site; however, according to Parrish, he stayed mostly on the sidelines, happy to let the workers do their work.[89] Also looking on were many of Bucky's university students and a horde of on-lookers (according to some reports, perhaps as many as 100 at its peak).[90]

At least one person who dropped by that day had a movie camera with them. Color footage of some of the day's building progress was captured on film; excerpts can be found on YouTube.[91]

Bucky was apparently pleased by the number of high school students who showed up to watch and ask questions during the day. "And not one of them asked if the dome could be brick-veneered," he later boasted.[92]

One of those who was in attendance was recently graduated SIU design student Herb Meyer. According to Meyer, before a single board was unpacked that day, there was already a buzz about this unique, about-to-be-raised home. "We just generally understood that it was going to be an event," he said. Parrish and his men were so focused and the work so efficient, there was not much to look at. Eventually, though, Meyer found himself put to work. Meyer got drafted to install some of the sealant tape that was being applied to the rooftop panels. Meyers says now, only half-joking, "Those strips were soaked in methyl-ethyl-ketone. They put me in charge of the toxic chemicals!"[93]

In contrast to the heavy-looking façade of brick or wood, the colors of Bucky's home would be calming and peaceful. The roof was a gleaming white. Meanwhile, its ten vertical surfaces (five trapezoidal and five triangular) around the base of the home were painted a soothing shade of "Ocean Blue." Then, interestingly, on each of the five wooden arches, or "canopies," over each set of sliding doors, these boards were painted a slightly lighter color of blue, creating a subtle effect of color gradation; it grows lighter in shade as the eye travels upwards towards the roof. A passionate sailor, it is speculated that Bucky chose the colors to mirror the white sails of boats gliding across the ocean.[94]

Once completed, the new dome home looked like nothing else in Carbondale—or many other places, for that matter. In 1964, *Time* magazine likened the house's appearance to a "pincushion without pins."[95]

The description is apt, but the home is not a perfect sphere. As wood is rarely (*per se*) curved, tops of geodesic domes (including Carbondale's Bucky Dome) are never

Workmen, students and others are pictured around—and on—the rough-in of the Dome.

Final rough-in for the Dome. Note that all the seams are now taped and covered. (*Courtesy: H. Meyer*)

completely round but made up of flat, triangular pieces that, when united, take on an "upside-down bowl" shape. Then, as mentioned, evenly spaced around the outside perimeter of the home are five vertical trapezoidal surfaces. Three of these surfaces are outfitted with sliding-glass "patio doors." One flat surface contains a "normal" traditional front door with "normal" adjacent awning windows positioned next to it. The final flat surface, in the home's bedroom, also has two awning windows, also stacked one above the other. During the Fullers' time in the Dome, the couple would cover these windows with venetian blinds.

Obviously, such an unusual home in such a "normal" neighborhood was destined to attract attention. Along with a steady series of drive-bys from curious Carbondale residents, in May 1960, a reporter from the Associated Press also came to write on the birth of the original dome home. According to the story, Bucky's Carbondale neighbors had already labeled the home the "Haystack House." The resulting AP story—reprinted nationwide—then brought with it a new batch of interested onlookers. Within a month, Bucky's secretary had received over fifty letters from thirty different states, all of them asking, "How do I find out more about the haystack house?" All letters were eventually turned over to Pease.[96]

Actually, for the people of Carbondale, in the early 1960s, it must have seemed like domes were taking over. Not only was the Fuller homestead erected in 1960, so was another geodesic dome. This one was on the corner of Southern Illinois and Grand Avenues. This dome was originally the home for a campus minister and his wife. After that, it housed the offices of Synergy, a walk-in treatment center. After that, the dome served as an interfaith center. In 1994, it was dismantled.[97]

Meanwhile, over on SIU campus, in 1961, four domes were raised by the Design Department to be used as classrooms; these domes stood until 1984.[98] Later, in 1962, Bucky was on site when a "huge" dome was constructed on SIU campus. Assembled by nine students, this dome, 72 feet in diameter, was a "Basketry Tensegrity Geodesic Dome" that stood five stories high.[99]

Also at SIUC's Campus Lake, there was once several partial domes (offering shelter but not closed in) that for many years provided space for students to study or just "hang out."[100] Of these partial domes, only one is still present (albeit completely replaced on its original piers). Today, this spot is known as "Bucky's Haven."

Then, in 1973, Illinois's DuQuoin State Fair got into the dome building action when they put up a partial, brown-shingled dome on its famous fairgrounds. For many years, that dome was the annual home to all SIU activities during the annual fair and served as a popular picnic spot during the rest of the year. In recent years, during the Fair, the dome has become a spot for local wineries to offer samples.[101]

Due to a geodesic's relatively easy assembly and relocation, over the years, SIU/Carbondale has also been home to many short-term domes. For example, in Betty Mitchell's 1993 *Southern Illinois University: A Pictorial History*, she reprints a photo of a group of Greek Row students who built a metal dome in front of their house and then moved it to the midway for Spring Festival 1963.[102]

Some years later, around 1975, another dome home was erected in Carbondale. This dome, about 2 miles from the "Bucky Dome," sits in a leafy neighborhood street on the city's Friedline Drive. Smaller than the Bucky Dome, the Friedline Drive dome is covered completely by white shingles; it is still standing today.[103]

The finished Dome with signs posted in front noting the contributions of Parrish Construction and of Pease Dome Co.

Bucky's Haven, an open-air dome, located at SIUC's Campus Lake in Carbondale.

"Dome fever" was spreading far beyond Carbondale. Lots of geodesics had begun to populate the globe by this time, popping up (as one Bucky-follower once put it) "like mushrooms after a rain." In 1959, near Cleveland, the American Society of Metals put up an open-lattice dome to serve as their headquarters. This still-standing dome, designed by Fuller, is 103 feet high and 250 feet in diameter. Additionally, construction on the Wood River Dome in Wood River, IL, started in 1959. Measuring 380 feet by 120 feet, and possessing 110,000 square feet of space underneath, this dome was a companion of sorts to the even larger Union Tank Car Dome build in 1958 in Baton Rouge, Louisiana. The Baton Rouge dome measured 384 feet across and 128 feet high; the structure served as a made-to-measure repair space for Union's railroad cars. That dome was utilized until new rail cars began being made too large for it. The Union dome was demolished in 2007. The Wood River dome, completed in 1960, however, is still standing.[104]

In January 1960, New York's Museum of Modern Art erected a "Geodesic Rigid Dome" in its garden space.[105] In 1966, when Fuller served as scholar in residence at San Jose State College in California, a 39-foot geodesic dome was erected in the middle of the campus.[106] In 1970, a geodesic was built in Antarctica to serve as the South Pole Station; this dome, 160 feet wide and 52 feet high, stood until 2003.[107]

Fuller was not done. Throughout the decade of the 1960s, Fuller announced plans to build domes and other major projects all over the world. At one time, there were plans for a Japanese baseball stadium dome, and in July 1966, Fuller was tapped to be the architect for the proposed "World Man Center" in the nation of Cyprus.[108, 109] In '66, Fuller suggested that the Illinois city of Herrin put a 300-foot dome over its central Park Avenue.[110] There was even a movement afoot at one time to have Fuller design a new city hall for the city of Carbondale.[111] Unfortunately, none of these proposed structures ever came to pass.

Equally audacious, but also so far unrealized, was Fuller's 1965 plan for urban renewal in Harlem. For it, Fuller envisioned the construction of fifteen cone-shaped skyscrapers, each 1,500 feet high, each so efficient, attractive, and reasonable in rent that their construction would reverse this area's ghettoization.[112]

In 1968, Fuller proposed a floating tetrahedronal city, 2.5 miles high, for the increasingly cramped people of Japan. The city could be floated out onto the ocean and then anchored in place.[113] Then in 1971, there was the announcement of Fuller's extraordinary plan for East St. Louis. Intended to be built in (over) East St. Louis, Fuller's proposed dome (stretching over a half a mile in diameter and reaching some 900 feet high) would be environmentally controlled and provide a home to 250,000 residents. The roof would span across land as well as over the Mississippi River. Like his earlier proposed silos for Harlem, this East St. Louis project was meant to counter the growing poverty of the area.[114]

The 1960s did however bring about two extraordinary, fully realized Fuller projects. The first had its seeds sown in 1965 when it was announced that Fuller was to design the U.S. pavilion for the forthcoming World's Fair that would take place in Montreal, Canada, in 1967.[115] The name of the Fair was "EXPO '67" and Fuller designed for it what many consider his greatest achievement—his sky high "Sky Break Bubble" mega dome.[116]

The finished building was a giant clear glass "ball" and was the largest transparent structure in the world at that time. It measured 250 feet in diameter and 187 feet tall and was nineteen stories high. It contained over 7 million cubic feet of display space spread out over multiple levels. The finished building is said to have come in under budget, with Fuller returning some of the original monies given to him by the U.S. government.[117]

The Expo dome—which Bucky nicknamed "Anne's Taj Mahal" in tribute to his wife—was opened on April 28, 1967. It soon became the talk of the Fair and the Fair's most iconic image. The majesty of the finished structure came to overshadow just about everything else about the EXPO, even the America-centric items it was created to celebrate. Though its outer skin was accidentally burned off in 1976, the skeleton of Fuller's Expo dome still stands and, today, it encases the Montreal Biosphere, an environmental museum.[118] Fuller's other great project of the 1960s was the development and launching of his "World Game."

One of the first steps for the World Game was when, in 1965, under Bucky's watchful eye, SIUC students constructed a 30-foot Dymaxion Map on the floor of a campus space. It took over four days and twenty pre-painted triangles to construct, but when done, the map would serve as the giant "game board."[119] World Game, as described by Fuller, emphasized design over politics, and "livingry" over "killingry." It was the opposite of war game simulations. It was a "mutual-success-seeking game" where the point was to have everyone win and nobody lose (and if someone lost, then everyone lost).[120] World Game was first played on the SIUC campus. From there, it spread to other colleges and, as an exercise, began to be "played" as part of corporate seminars. It was even discussed in Congressional hearings.[121]

As for the Fuller home now standing in Carbondale, to stand and look at it, it can be a bit misleading; the Dome is deceptive. Though the home might look small from the outside, thanks to the lack of straight lines and the high ceilings characteristic of a dome, the interior feels much larger than it appears. More than one person over the years has compared it to the TARDIS from *Doctor Who*.

Yet, even then, it is still a one-bedroom home and some of Bucky's colleagues were often surprised when they visited and learned that someone as renowned as Fuller did not live in something more luxurious. Despite having once described himself as a "reformed millionaire," Fuller was not, at this point, interested in ostentatious displays.[122] He preferred the practical, and with their only child, Allegra, now out on her own, the Fullers had no use for an expansive amount of space. Furthermore, the year the house was built—1960—was long before the "McMansion" movement of later decades. Hence, this for them, at this time, was a perfectly sufficient abode.

In such a circular home, especially one equipped with several sliding patio doors, it is debatable about which door actually constitutes the "front" door. Nevertheless, on the Dome's northeast-facing vertical trapezoid, there is a single-sided entrance that serves as the house's "main" entryway. Since it is adjacent to the driveway and contains the only "swinging" exterior door at the house, it is the logical choice for a main entrance.

In Bucky's day, the front door consisted of a full glass aluminum storm door in front of a wooden door; the wooden door was set with a panel of glass almost the door's entire height and width. Positioned next to the door are two rectangular windows, each measuring, roughly, 2 feet high and arranged one above the other. The total window, then, is 4 feet by 4 feet 6 inches wide. Likewise, the windows in the north-side

The finished Dome. Note the fence that surrounds the property.

bedroom window are identical to the entryway windows. Since each window has the two stacked panes, they can be each be opened outward in an awning fashion. This provides important air circulation while also offering some protection from precipitation during humid Southern Illinois summers. These two sets of windows are the only non-sliding doors and windows in the home, not counting the fixed skylights in the roof.

In their plans, Pease did not dictate how the insides of their domes were to look; this was left to the purchaser. It is believed that Bucky deferred to Anne to create much of the layout of the home's interior, possibly with the assistance of her friend, Robert Hunter.[123] What the duo came up with was a straightforwardly elegant, yet efficient, design—one where obvious thought was put into the room flows and the placement of closets and other amenities but which also, notably, echoed that of Fuller's earlier Dymaxion home.

If you enter the Carbondale Dome through the front "entry" door, you find yourself in the house's small foyer. There, the visitor is immediately faced with a choice of left, right, or forward.

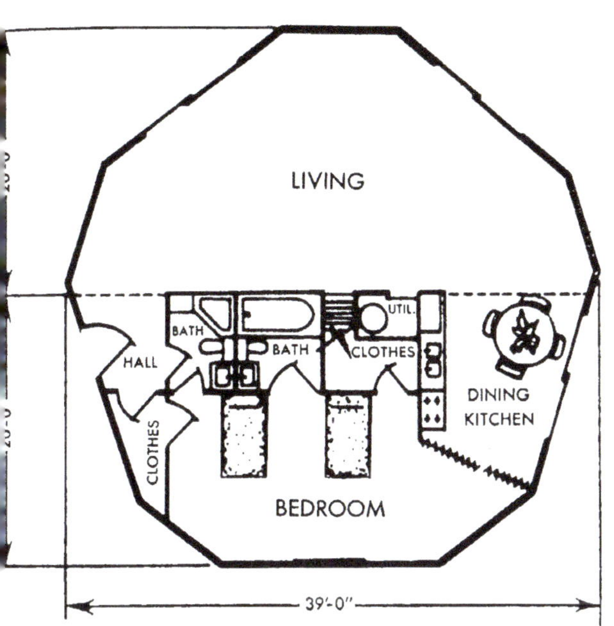

Above left: The Dome's interior, floorplan for its first floor.

Above right: The Fullers' front door pictured from the inside.

Choosing left would take one first into the home's main living space, its vaulted living room that constitutes about half of the home's floor space. Two large patio doors as well as four of the structure's overhead skylights make the space very well lit and amplify the interior's spacious feeling.

Continuing on the house's main floor, the living room flows seamlessly into the dining room and around to the kitchen. Against the kitchen's interior wall is the home's double-bowl sink, wood cabinets (upper and lower), and major appliances (Hotpoint refrigerator and stove and Fasco range hood). The kitchen countertop was white with gold flecks and this same material and pattern was reproduced for the house's backsplash. Another sliding patio door is present in the kitchen, allowing yet another way to exit the house while also admitting sunlight. Additional light is added in the kitchen—and in the bedroom—via several overhead, inlaid light fixtures (two in the kitchen and three in the bedroom), flush with the ceiling. Intentional or not, the flat, square appearance of the lights mimic the look of the Dome's skylights.

Following further around the circumference of the home, through a hanging, vinyl-faced door that folds accordion-like against the wall (another Fuller characteristic), one enters into the home's bedroom. In the Fullers' day, the bedroom was distinguished with two twin beds, a layout that mirrors the design of the old Dymaxion home. At the back of the bedroom is access to the home's master bathroom. There is also located there a single closet with double doors. The modest closet is equipped with both a rod

Bucky and Anne at home. The hanging sculpture behind Anne is by Ruth Asawa.

Anne Fuller pictured in the Dome's kitchen. (*Courtesy: Department of Special Collections, Stanford University Libraries and the Estate of R. Buckminster Fuller*)

Above: The Fullers' kitchen; the same stove and range top are still in the home today. Note the pencil sharpener affixed to the wall at extreme left.

Right: The kitchen leads into the bedroom. Note the accordion door. The two photos visible on the wall in the bedroom are believed to be of the Fullers' two daughters, Allegra and the late Alexandra.

for hanging clothes and a series of wooden shelves. The bedroom was also the location of one of the home's telephone jacks; the second was up in the loft.

Exiting the bedroom, at the opposite end from where one enters, takes one through a (very small) vestibule and storage area where built-in shelves, cut to fit the geometric shape of the Dome, attempt to make the most of this (perhaps leftover) space. Additionally, cut into original wood of one of the shelves, and then again directly across from it, were two notches into which a horizontal pole could be placed. Assumedly, this was created as a hanging rack to drip-dry clothes. The Dome has never been equipped with a washer or dryer.

Finally, following the home's normal, circular flow, one arrives back to the small foyer.

There is also a door located in the foyer (directly in front of the home's front door) that leads into the home's other bath—a "guest" bathroom with a shower. This second bath is also accessible from the bedroom and was for Bucky's use. That accessibility is a unique feature that Fuller drew into the original house's design; it granted him and Anne separate his-and-her bathrooms. In this second bath, a linen closet was built into the wall next to the shower stall. The bathrooms have many similar features but, interestingly, different vanities. In "her" master bath, the vanity has a more glamourous, shiny, reflective black finish and silver hardware; the vanity in "his" bathroom is plain white and simply serviceable.

Concluding the tour, in the living room, found just off the center point of the home, is the staircase that leads up to Bucky's second floor study, a highlight of which is the specially-built curved bookcase that runs along the room's/home's upper back wall and, again, reflects the geometry of a round home. Bucky would later equip that upper room with a chair on wheels allowing him to roll over to any book with ease. The Dome's other six skylights filter into the study.

In his 1977 book about Fuller, *Cosmic Fishing*, writer E. J. Applewhite said about the home:

> The house was a microcosm of Fuller's universe: spherically coordinate, uncompromisingly simple in design, and at home in its environment. Its scale and weather-wood framing were quite in harmony with the conventional house with front porches and side yards that comprised the rest of the shaded neighborhood. As you enter the house the first impression is the absences of the familiar four-squared cubical framework of rectangular floors and straight walls. The effect is totally disorienting to our reflexive assumption that rooms should be shaped more or less like shoe boxes.[124]

Though the home itself rose from the ground quickly—essentially in one day—completing the interior reportedly took six months as plumbers (from the area's B&C Plumbing and Heating), electricians (from local Anderson electric), and carpenters figured out how to do their work in a home with very few right angles. Due to the ongoing interior work and finishes, the Fullers were probably not living in the home until November 1960.[125]

Of course, once built, plumbed, wired, and made inhabitable, the Fullers commenced decorating their house, turning their dome into a home. Throughout their residency, the home's interior décor would, of course, change and evolve, but it would always remain modern and a little eclectic—much like the home's two occupants.

Against a backdrop of its pristine white interior walls, most of the home was originally filled with many of Anne's most beloved and unique finds. Mrs. Fuller was a passionate collector of antiques and possessed a fine eye. The home would prove to be a wonderful showcase for her treasures. Anne worked closely with her friend, Sally Roan, whose

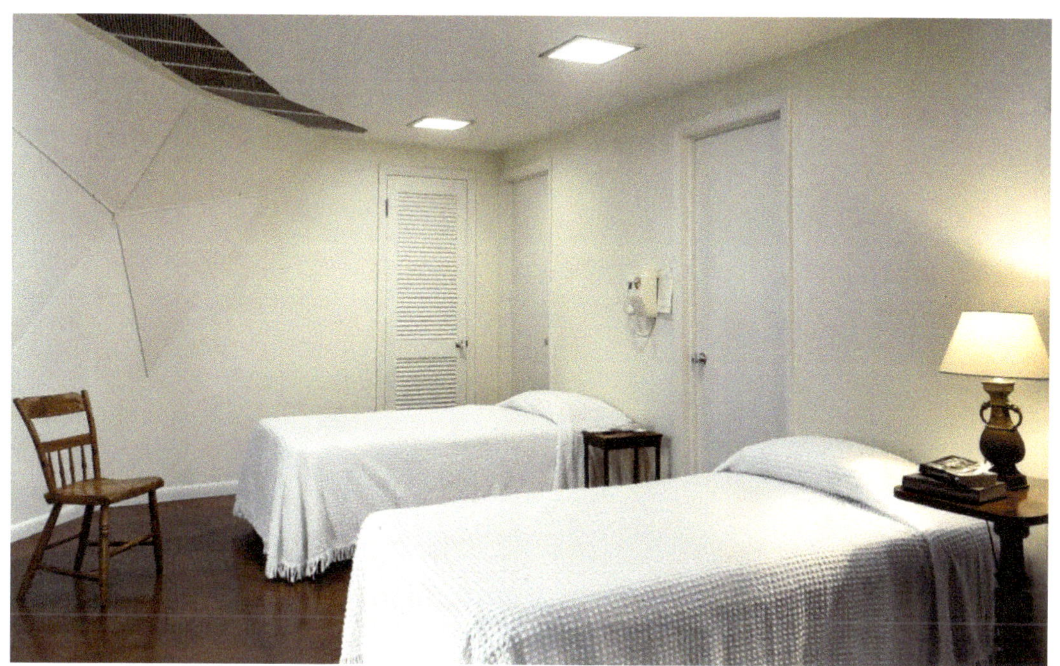

Above: The Fullers' bedroom. The doors at right and center are to the home's two bathrooms; the door at the far left leads out of the room. Note the telephone positioned on the wall between the two beds; note also the vents in the ceiling to allow for airflow.

Below left: Shot of the living room and the stairs leading up to the loft. Wooden planks serve as guardrails for the second floor. Note the industrial fan that hangs from the center of the home.

Below right: The loft—always a passionate sailor, a ship's model of Bucky's is in the foreground of the photo. (*Courtesy: Department of Special Collections, Stanford University Libraries and the Estate of R. Buckminster Fuller*)

A shot of the living room furnishings; the woman in the photo on the table is the Fullers' daughter, Allegra.

husband, Herb, taught alongside Bucky at SIUC, to make the space look and feel perfect. With time, the home would also become decorated with items, from places like Java and Africa, that Bucky had collected or been gifted with during many of his travels.[126] From photographs and other sources, we know that the house included a modern-looking white leather sofa; a comfy "bucket" chair; a couple of antique wooden "secretaries"; a couple of floor lamps to augment the home's natural lighting; a ladder-back chair (from Anne's great-aunt Margaret); two large ceramic, Chinese Fu Dogs (hundreds of years old; originally owned by Anne's grandfather); an old barometer; a new telescope; a Japanese electric clock with an airplane sweep hand; an African-carved stool; a large metal sculpture by Ruth Asawa; and various exotic clay pots from various far-off lands. One of Bucky's favorite pieces of furniture in the home was a three-legged chair (a fourteenth-century antique) given to the Fullers as a wedding gift; Bucky, apparently, was quite fond of the chair's triangular shape.[127]

One of the home's most interesting bits of décor was the Fullers' replica of the "Lincoln Imp." This gargoyle-like statue of about 8 inches was based upon an English legend and was rumored to bring good luck; it hung in an upper corner, at a vertex, on the home's western wall (or ceiling, depending on how you look at it).[128]

Throughout the first floor, wall-to-wall carpeting was rejected in favor the home's cork flooring, which was originally believed to be wax coated and buffed to a shine. Later, large Oriental rugs were introduced to soften the space. The lack of flat, vertical wall space (for the most part) prevented the hanging of traditional pictures and paintings;

Above left: Another shot of the Fullers' living room with some of their early furnishings.

Above right: The Fullers' Lincoln Imp.

"They would just sort of dangle out from the curve," Anne once lamented.[129] Instead, the Fullers utilized standing columns or easels or they suspended objects—including a Japanese paper lantern and, fittingly, a wire model geodesic sphere—from some of the ceiling's upper beams or "chords." Two large Japanese woodcuts—one representing "Day," the other representing "Night"—were showcased by placing them on the front of the loft's upstairs rails.[130]

In their dining room, the couple utilized a solidly built Queen Anne's table around which they gathered four Eames chairs. It was at this set-up that Bucky wrote many of his books. Bucky said once about his interior design philosophy, "Anything that is exquisitely well designed of any period harmonizes with things that are equally well-designed no matter when they are crafted."[131]

In the bedroom, the couple placed on their two beds matching white chenille bedspreads. Matching bed tables stood beside each. On Anne's side of the room, the side closest to the closet, she placed a large dressing mirror, its edges framed by thick, dark wood. A small writing desk sat pressed against the north wall. Later, on the bedroom's walls, the couple hung two portraits—one each of their daughters, the late Alexandra and one of Allegra.[132]

A newspaper record from the era described the home as "bright, cheerful and amazingly spacious." That same article quoted Anne as saying the home was easy to keep clean.[133]

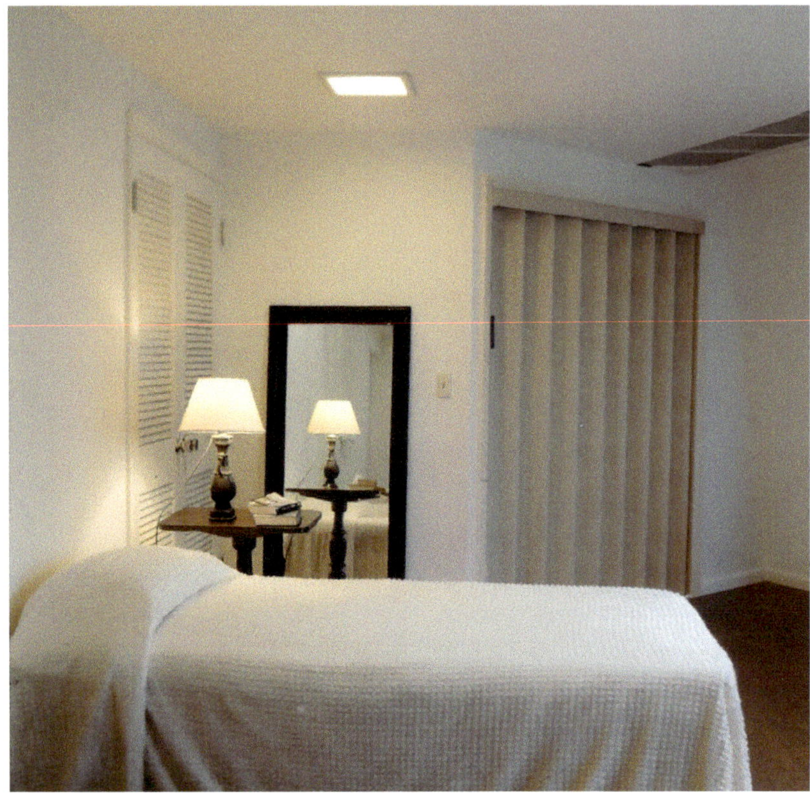

Anne's corner of the bedroom.

Based on photos, the Fullers' completed the exterior landscaping of their home with a perimeter sidewalk (an outgrowth of the home's concrete foundation) that extended about three feet out from the outer edge of the home. Additionally, an expansive "apron" of white gravel was also placed outside the house's south side sliding doors creating a patio effect. Two outdoor lounge chairs complimented the outside décor. In time, the Fullers would add two or three saplings, in wooden pots, that would sit up by the house. Later still, the Fullers would also add a small, outdoor sculpture.[134] Sadly, this artwork would be stolen in December 1967. Wanting it back, the Fullers placed an ad in the newspaper and offered a reward. Alas, there would be no recovery.[135]

Anne said about the Dome and its surrounding landscape:

[It] was very nice because we had enormous windows around the whole circle. We had the garden with a fence you could not see through though the air could come through, and there were lots of trees that screened and shaded us. So we used to be outdoors a lot.[136]

Up in his study, Bucky had a large record player nestled into its own four-foot by two-foot cabinet. With the volume up, music filled the house. If the sliding doors of the home were open, the sound would carry outdoors and the Fullers and their guests could enjoy music as they sipped tea in the garden, enjoyed Anne's flowers, and breathed in the scent of the yard's mimosa and poplar trees.

Once again, though, just how much time Bucky actually spent in the Dome during his Carbondale years is debatable. Due to his active lecturing schedule and his two-months-a-year agreement with SIU, Bucky was often out of town. Sometimes, due to his frequent absences, he referred to the Dome as his "private motel." It was a term he first used in 1960, practically as soon as the house was built.[137]

Regardless of Bucky's absenteeism, Anne was frequently at the house. An old photograph of Anne walking from the Dome carrying out the trash illustrates that while her home was an architectural innovation, it was also just home.

Though Mrs. Fuller is, unfailingly, remembered as the height of graciousness among those who knew her, during her Carbondale years, there remains some speculation surrounding Anne's overall happiness. With her husband often away, and with no family close by, it is quite possible that Mrs. Fuller did, at times, feel rather alone or isolated.[138]

Additionally, Mrs. Fuller no doubt found herself in for some culture shock when she moved to Carbondale. SIUC's student population at that time was a fraction of what it would become in later decades. With the campus so much smaller, so was the city. Having grown up in more populated areas, always with easy access to big urban centers and cultural landmarks, Bucky's wife probably felt like the reverse of *The Wonderful Wizard of Oz's* Dorothy; instead of suddenly finding herself in a new, exotic place, she,

Exterior of Dome with the fountain in the foreground.

Above: Anne Fuller at the back of the Dome. Note near the center of the photo, one of the two lamps jutting up from the cement stairs that illuminated the steps at night.

Left: A beaming Anne Fuller, photographed by her friend Michael Mitchell. (*Courtesy: M. Mitchell*)

instead, found herself suddenly surrounded by corn fields. Thankfully, many of Bucky's staff and students and colleagues often circled around Mrs. Fuller, becoming a local, *de facto* family for her.

When Bucky was at home, he and Anne enjoyed entertaining in their unique home. Evening visits by friends often included a rendition of Bucky singing his specially written song, "Roam Home to a Dome," sung to the tune of "Home on the Range."[139] Fuller was so fond of his tune he once arranged to make a recording of it with him singing it and accompanying himself on guitar. It was recorded at the broadcasting studios of SIUC's Department of Radio-Television in 1956. The original tape, long thought lost, was discovered again in 1988 and donated to SIU's Morris Library.[140]

Once, for a few nights, the Fullers welcomed architect Thomas Zung, who remembers:

> I had the pleasure of sleeping once up in the library on an old WWII army cot. I prayed the canvas would not split and crash, disturbing Anne downstairs. I did not have to worry about Bucky, as he took out his hearing aids at night.[141]

Besides Zung, Mr. and Mrs. Fuller were also often visited by Michael Ben-Eli. An architectural student in his native London when he first met Fuller in 1964, Ben-Eli was struck by many things Bucky had to say including Fuller's desire to "redesign the world" with architecture leading—not following—social change.[142]

After Ben-Eli graduated, in 1969, he joined Bucky in Carbondale. The two men originally met up in Maine before traveling to Illinois where Ben-Eli joined Fuller's campus staff which included Shirley Sharkey (*née* Swansen), Mike Paterra, Mark Victor Hansen (who later went on to co-author the very successful *Chicken Soup for the Soul* book series), and Max Ackerman.[143]

Ben-Eli traveled with Bucky often, but during the times Ben-Eli was in Carbondale while Bucky was out of town, he would often assist Anne with her weekly marketing or just join her for an evening cocktail. He says, "I liked Anne very much. She was quiet but tough-minded, in a nice way. Gentle and generous—a real lady."[144] Today, Ben-Eli heads the Sustainability Laboratory and remains associated with the Buckminster Fuller Institute.[145]

Two other visitors to the Dome were architect George Anselvicius and his fiber artist wife, Evelyn. Their visit to the Dome was documented by local newspaper photographer Ben Gelman.[146]

In 1965, Fuller was visited at his Dome by the FBI. The government agency had been following Bucky's activities for several years, apparently alerted by an unknown SIUC-based source to Bucky's occasional trips to Russia and Fuller's once announced hoped-for trip to Cuba. Things were resolved quickly however.[147]

Of course, even among so many distinguished guests, the Fullers' favorite visitors were their two grandchildren. Though grandson Jamie Snyder was probably only about seven the first time he visited his grandparents in Carbondale, he nevertheless remembers how the trip came about. One springtime, his parents gave him a choice: they were going to Europe and Jamie could travel with them on the ocean voyage or he could go stay with Grandma and Grandpa in Carbondale. Jamie chose the latter.[148] Snyder recalls:

> I had a close relationship with my Grandpa. He was a great teacher and he loved having another inquisitive mind around. I remember their house as a beautiful place,

Above: Fuller with students in the yard in front of the Dome. (*Courtesy: B. Ashley*)

Left: The Fullers in their dining room joined by their friends George and Evelyn Anselvicius.

nice. I loved the Dome. It had a nice feeling to it. Some of my memories are fleeting but I remember the cork floors—the heating in the floor. The floor was very tactile, a soft cushion.[149]

Other memories of Snyder's are equally impressionistic: "I remember going for a bike ride with him. Riding from the Dome. I remember driving with him. He drove a black Citroen sedan. It looked a bit like the Dymaxion car. He loved driving it."[150] Bucky also owned a Cushman scooter, which he used to travel the short distance from his home to campus, and he also owned one of the first electric cars ever sold—a Henney Kilowatt.[151]

Jamie's slightly older sister, Alexandra, once joined her little brother for a later visit to their grandparents' home. Both grandkids remember a trip their grandfather took them out onto the water. Jamie relates, "He had this fiberglass dingy, a cat boat, that we launched on a lake. Of course, Bucky loved sailing and I remember it being very exciting."[152]

Another time, during this same visit, Grandpa Bucky helped his grandson learn to tie his shoes by moving the shoelaces from his actual footwear and onto the handles of one of the cabinets in the kitchen. Jamie says, "He was always looking for innovative, new ways of doing things and, somehow, by changing my orientation for my shoe-tying, it helped me grasp it quicker."[153]

Today, Snyder is unsure about the length of his first visit to his grandparents' home. Was it a matter of weeks or a month? He believes he slept on the living room sofa during his stay in the Dome before traveling with Bucky and Anne out to the East Coast, ultimately, to rejoin his parents.[154]

The Fullers' driveway: parked in the rear is the couple's Citroen; in the foreground, partially cut off, is their Henney Kilowatt.

Alexandra visited her grandparents in Southern Illinois only that one time. Nevertheless, their home made an impact on her. She says today, "I liked the space; it had a nice feel to it. My grandmother loved creating a home. She had a very modern design aesthetic but it felt good to be there. It wasn't quaint, but it was cozy."[155]

Some years later, around perhaps 1967, a now teenaged Jamie returned to Carbondale. This time he arrived alongside his father, filmmaker Robert Snyder, to make a documentary on Bucky. By this time, Jamie was involved passionately in his hobby of photography and was brought onto the production team to work as its still photographer.[156]

The finished special was shown in February 1971 as *NBC Experiment in Television: Buckminster Fuller on Spaceship Earth*. Production of the program began with a stop in Carbondale and filming at the Fuller house. After one or two days there, the film crew went on to the University of Detroit.[157]

Not surprisingly, domes and other Fuller innovations dominated much of the grandkids' growing-up years. As kids, in California, where they were raised, the family had a play dome in the backyard. Later, Jamie recalls once building a geodesic dome for a junior high math class—and having his grandfather help him do it.[158]

Jamie says, "He was a wonderful grandfather. He knew how to sail, he had a cool car, he knew how to build models and he knew geometry and photography."[159] Though, even at a young age, Jamie realized that his grandparents' abode was out of sync with other homes, to have his grandma and grandpa reside in a dome "just felt right, [it] just felt like grandma and grandpa's house."[160]

Despite its many advantages, sometimes, for the Fullers, day-to-day life inside the Dome was not easy. Contrary to what Bucky expected and Pease promised, the sealant for the roof did not prove completely waterproof, and interior leaks were a sometime problem.[161]

However, in true Fuller fashion, Bucky treated this weakness as an opportunity to experiment. After the initial Celastic tape application failed—it did not expand and contract in weather and in the sun as, ideally, it should have—Bucky next turned to rubberized, or plasticized, paint to seal up the topside surfaces. After that, too, proved unsuccessful, Fuller applied a special urethane spray-on foam. That, too, however, proved faulty. Finally, after becoming convinced that the type of epoxy he needed to use had not yet been invented, Bucky relented to covering his house with normal, old-fashioned, but functional, asphalt shingles. It was a "fix" that would endure for many years.[162]

Though the addition of the shingles kept most of the leaks at bay, it dramatically altered the look of the Dome. Gone was the bright white roof simulating windswept sails; now, the Dome had a rough and greyish tint to it. In some areas on the Dome, the shingles extended all the way down the sides of the home covering up much of the pretty ocean blue. In the end, only the blue around each of the doors remained visible.[163]

In early 1968, the Fullers' outer fence was damaged. On the morning of January 22, a local businessman was driving down Forest Avenue when he suffered a fatal heart attack. His vehicle careened into the Fullers' property and through their fence. The gentleman was pronounced dead at the scene. Neither of the Fullers were at home at the time of the incident.[164]

Even with these issues, though, Bucky and Anne made the most of life in Carbondale and had a small group of kind neighbors who enriched their time while living in the "land of the Saluki."

Bucky poses in front of the home. Note that the roof—and some sides—of the home is now covered with shingles.

Already living in the neighborhood before the Fullers arrived was Robert Childs and his wife, Francis, who had lived at 806 West Cherry since 1958. Today, Mr. Childs admits that, in 1960, he was not familiar with Buckminster Fuller but was "pleasantly intrigued" when the Dome arose on that April morning. As Mr. Childs recalls, the house was not there when he left for work but was there by the time he got home. The Childs were even more pleased that the owners of the Dome turned out to be such good neighbors. Bucky, Childs says, was "easily thirty years ahead of his time ... Thankfully, he spoke in a language I could understand."[165]

The Fullers were also generous. Mr. Childs, a member of the Episcopal Church, recalls:

I visited at length in the Dome with the designer in quest of a pledge during the 1965 Every Member Canvas.... I was surprised when Bucky said he and his wife, Anne, although strong Episcopalians, could not make a monthly pledge because he had no regular income. He would, however, give me a lump sum if that would suffice. I indicated it would be most satisfactory, and Bucky made out a check I placed in my canvassing envelope. Later examination showed it to be for $2,000—real money in 1965.[166]

Another neighbor, Ann Dillard, was also already in the neighborhood when the "weird" house went up, vividly recalls the rampant discussion in the neighborhood that followed immediately after. Unfortunately, the imposing wraparound fence that gave the Fullers privacy also served as something of a social barrier. Mrs. Dillard remembers seeing the Fullers "only a couple of times" around the neighborhood before she and her family moved in the early '70s.[167]

Neighbor Shirley Maine concurs about the impact of the tall, imposing barrier around the property. She says, "The Fullers were kind of cut off in some ways because of that high fence."[168]

Maine was not there when the Dome went up; she did not come to the neighborhood until 1963. At that time, she admits she also wasn't completely familiar with Buckminster Fuller and would remain that way until years later when she saw a play about him in Chicago. Additionally, with Bucky often traveling, Maine did not see too much of him but became friendly with the "sweet" Mrs. Fuller through her son, who was the neighborhood paperboy and had the job of mowing the Fullers' lawn. Maine's son, Patrick Lillard, also remembers the Fuller home as the "go-to" house on Halloween; Mrs. Fuller always made a habit of giving out full-size—as opposed to "bite-size"— candy bars to all the trick-or-treaters.[169]

The summers that Lillard did not mow the Fullers' yard, it fell to young Robbie Stokes whose family also lived nearby. Though Stokes does not remember how he got hired to cut the "postage stamp-size" yard, he remembers getting paid 50 cents every time he did it and that Mrs. Fuller always brought out lemonade for him afterward. He also distinctly remembers one time when Mr. Fuller met him in the yard to show him an object he had in recently acquired. Stokes says, "It was an ancient Israeli oil lamp which Fuller said was around 2,500 years old. He showed it to me."[170]

Another of the Fullers' one-time lawn-cutters was Gary Wallace. Wallace was the stepson of Bucky's long-serving secretary Naomi Smith Wallace. Wallace, now an SIUC grad living in the Chicago area, remembers doing lawn work "for about three years" at the Bucky Dome in the late 1960s. The Fullers paid him $20 every time he cut the grass. Wallace says now, "I thought I was rich!"[171]

Mrs. Fuller was also quite thoughtful towards the Poppen kids, who lived a few doors down. Siblings Lisa and Dylan had lived on the block with their parents beginning in 1971, near the end of the Fullers' tenure in Illinois. Nevertheless, the kids were there long enough to make the acquaintance of the "friendly" Mrs. Fuller and for her to sometimes babysit Lisa and to send back to both kids postcards from locations like Japan and Alaska. Sometimes, Mrs. Fuller even brought them back small presents from her travels. She would often regale them with stories from her wide-ranging adventures. Though the Poppen kids were quite young—still toddlers—Lisa was nevertheless quite aware of having such a "famous man living right next door."[172]

Yet if Lisa was sometimes kindly babysat by Mrs. Fuller, her older sibling, Dylan, sometimes drew the (gentle) ire of Mr. Fuller, as Dylan recalls:

My friend Mitchell and I would sometime climb up on top of the Dome. You could use the side door to climb up. We were six or seven; we were these "free-range" kids. It was a great roof to be on top of. Then Bucky would come out and tell us to get down. He'd hear us up there. He would come out and yell, "I'm trying to write!" Sometimes he'd throw these tiny pebbles or bits of moss as us. Nothing to hurt us. That's why he wanted us down, so we wouldn't hurt ourselves.[173]

Despite that reaction, however, Dylan remembers Mr. Fuller as being very "grandfatherly" towards the neighborhood kids, adding, "He was very straight-forward; he didn't treat us like kids."[174]

Later, not long before the Fullers moved out of Carbondale, Dylan remembers them having a yard sale. Dylan still owns the small, grey-colored, red-interior suitcase that he bought at the sale for a few cents. He says, "It originally had a tag on it that said 'Anchorage, Alaska/B. Fuller.' I think my sister has it now."[175]

Two other youngsters who have fond memories of the Fullers are Toni and Barbara Roan. Their father, Herb Roan, was a friend of Bucky's and a fellow professor. The families often visited each other. Toni Roan says:

> Bucky was like an uncle to me. And a fun uncle. He was fairly little with these thick glasses, very jovial and he said funny things ... I remember one time—I think I was about 10—my dad and I drove with him to the airport. He had a Citroen. It was so cool! I felt like I was riding in spaceship! It was like riding on air, so comfortable. All the way, Bucky tried to explain to me how it worked but most of it just went over my head.[176]

Some years later, the Fullers were at the Roan's when Bucky learned of Barbara studying dance. The next thing she knew, she was "up and dancing with Bucky, I think we fox-trotted or waltzed."[177]

Toni's older sister Barbara also remembers things associated with the Fullers. Once she was on campus and wandered into the second floor of Morris Library where Bucky's papers had just been moved. She says, "I remember large cardboard boxes were being used as room dividers and there was Bucky in the middle of it lecturing to a group. I snuck in and listened but I couldn't follow everything he said."[178]

Both girls were intrigued by the Bucky Dome, which they often visited. Toni says, "Anne was very gracious. A wonderful hostess. I loved [the house]. It was like having your own private jungle gym."[179]

Barbara concurs: "I liked how one room just flowed into the next. Though I do remember thinking that that square box right in the middle was a little strange."[180]

The Roans kept up with the Fullers long after they all departed from Southern Illinois. Toni received a wedding gift from Bucky and Anne when she got married in 1983.[181]

As can be imagined, with such a renowned couple living down the street in such a remarkable looking house, life in the neighborhood could be interesting. With time, though, the neighborhood got used to a certain level of media presence—local or otherwise—as various people came to talk to Bucky.

In January 1964, *Time* magazine put Fuller on their cover. Inside was a multi-page article recounting Fuller's youth and early engineering feats as well his many ambitious plans, like the proposed Manhattan project. The writer of the article visited Fuller in Carbondale and the article featured a photo of Fuller, Anne, and guests inside the Dome.[182]

It was also an interesting day when, in 1966, Walter Cronkite showed up. Arriving there to interview Bucky for a TV series titled *The 21st Century: Cities of the Future*, Cronkite first went door to door in the neighborhood asking folks for their opinions of Bucky. No one, it seems, had anything bad to say.[183]

Walter Cronkite interviewed Fuller in the Dome in 1966.

Fuller was good (and frequent) "copy" during his Illinois era. Along with the cover of *Time*, during his SIU years, Fuller would also be featured in *Newsweek* (1963); *The New Yorker*, *Saturday Review*, *Look* and *Popular Science* (all 1966); *The New York Times* and *Horizon* (both 1967); and *American Scholar* and *Reader's Digest* (both 1969).[184]

It was not just the printed press that loved Bucky during the 1960s. In 1961, Fuller appeared as a guest on the PBS talk shows *Spectrum* and *Off-Ramp*. In 1963, Fuller was profiled in the PBS series *The Fuller World*. In 1965, Fuller was twice a guest on Merv Griffin's talk show. On the same shows with Bucky were Zsa Zsa Gabor, Eydie Gorme, Ray Bolger, Xavier Cugat, and Charo. Fuller was on *Dick Cavett* in 1968 and David Frost's show in 1969. In July 1969, Fuller was a commentator on CBS-TV's Apollo landing coverage, and after that, he appeared on a TV forum hosted by Mike Wallace titled *CBS Special: Apollo*.[185]

Then Fuller went "behind the scenes" in TV when he was asked by a producer to serve as a consultant for the made-for-TV film *Earth II*. The 1971 film was set aboard a U.S. space station orbiting above the Earth. The film made plentiful use of Fuller's Dymaxion Map. In the film's closing credits, Fuller is credited as the film's "Technical Advisor for Earth, and Designer of Dymaxion Map."[186]

R. Buckminster Fuller in the loft/study of his Carbondale home.

Buckminster Fuller once described living in the Dome as "an interesting experience—getting up and going out to the University in the morning from this round house. As I go from the bedroom through the kitchen and dining area, and then out through the living room and finally all the way around again out the driveway, it's like coming out of a conch shell."[187]

The overarching symmetry of the home may have, in the end, aided in Fuller's thought process. His daughter, Allegra, has reflected, "[The Carbondale years] were an extraordinarily productive time for my father. Many important directions that became even more significant in his work began to evolve here."[188] It was, after all, during his time dwelling in Carbondale that the technology of the geodesic dome truly exploded, practically taking over the globe.

If further proof is needed of Fuller's rich Carbondale years, consider the impressive number of patents he was awarded by the Patent Office during this period. Beginning in 1959, with the "Catenary," or geodesic tent, Fuller would average one patent almost every year he resided in Carbondale. In 1961, he patented his octet truss design; in 1962, his tensegrity structures; in March 1963, he patented a "submarine island"; in 1964, he got a patent for a new suspension building system ("aspension"); and, in 1965, he was awarded three different patents: for the "octa spinner"; for the "monohex dome" (commonly known as the "fly-eye dome") and one for the "laminar dome." In 1970, Fuller received a patent for a rowboat he called Rowing Needles.[189]

During his SIU years, Fuller would also be showered with accolades. In 1961, he was named by Harvard as their Charles Eliot Norton Chair of Poetry. In early 1963, Fuller went to the University of Colorado for "Fuller Recognition Day"; in April, he was in London to be feted at a dinner with members of the British royal family; in May 1963, he became the first SIUC faculty member ever inducted into the National Institute of Arts and Letters. In 1964, Fuller received a "Doctor of Letters" from Clemson College and an award from Brandeis University. Other honors that arrived in 1965 came from the University of New Mexico and the Delta Phi Delta fraternity at Ball State College. In April 1968, Fuller was recognized as Boss of the Year by the Carbondale chapter of the National Secretaries Association. That same year, Fuller was also recognized by Southern Illinois, Inc. with their annual Area Appreciation Award.[190] Upon news of the award, Fuller said, "When I lived in New England, I had a small backyard, which grew bigger as I traveled around the world. Now when people ask me where I live, I say 'on a small space-ship called Earth.' And I am proud that Southern Illinois is the center of the backyard."[191]

Also in 1968, the Fullers traveled to England so that Bucky could receive the Royal Gold Medal for Architecture from Queen Elizabeth II. In 1970, Fuller received the Gold Medal from the American Institute of Architects. From 1963 to 1968, Fuller was an advisor to NASA, and, twice, Bucky and Anne were invited to the White House as guests of President Johnson.[192]

Some years after the fact, one-time head of the Department of Design, Bill Perk, stated that he believed this unusually fruitful period for Fuller was due, in part, to his surroundings. Perk said, "[Bucky] would point out that this rural setting allowed him to get things done that you couldn't in an urban area."[193] Long-time Fuller associate, Thomas T. K. Zung states, "Bucky had several golden years as many geniuses have. But, to me, SIU is one of the big golden eras."[194]

2

LIMBO (1973–1999)

limbo: lim·bo (/ˈlimbō/) n. an intermediate state or condition.

Buckminster Fuller would remain with Southern Illinois University at Carbondale until 1970 when a variety of university events led to his departure.

The era's youth-led political strife that was rampant in the country, especially on college campuses, was strongly felt in Carbondale. Against the backdrop of the Vietnam War, issues of racism, sexism, misuse of school funds, and other hot-button topics gripped and, in some cases, shut down Southern Illinois University. Student protests that year took over the campus and often wreaked havoc. The furor had been building for some time and reached a crescendo in 1969 when SIUC's most legendary building, Old Main, was burned to the ground during one protest. Then, in 1970, amid continuing scandals, the school's legendary, long-serving president, Delyte Morris, was deposed, more or less forced into retirement.[1]

Morris had long been Bucky's biggest advocate, and with his resignation, Fuller lost his greatest supporter. In addition to Morris' departure, two of Fuller's closest on-campus compatriots—chief counsel John Rendleman and Dean Bernard Shryock—both left SIUC around this time; Shryock moved into retirement, while Rendleman took over as head of Southern Illinois University Edwardsville.[2]

Additionally, at the beginning of the 1970s, Fuller's budget was cut by the university and his office also came under fire for various reasons and from various still-standing university factions. Sometime in 1971, things came to a head between Bucky and his main university liaison, Tom Turner, possibly over matters related to Fuller's budgeting and accounting practices.[3] There is also some speculation—though not fully substantiated—that Fuller grew upset when the school decided to build a new basketball stadium around this time and passed on the idea of covering it with a geodesic dome.[4]

Finally, there were issues related to Fuller's gigantic Chronofile, which was then housed at SIUC's Morris Library. For Bucky, the chance to finally find a place for his Chronofile—his staggeringly large collection of papers (dating back to his birth announcement), drawings, writings, and correspondence—was one of the reasons he

The Dome's second owner, Michael ("Mickey") Mitchell (left) pictured with Bucky and John Christy. (*Courtesy: M. Mitchell*)

came to SIUC in the first place. Since its inception, Fuller had, according to one of his biographers, Hsaio-Yun Chu, depended on his Chronofile for both "emotional support and academic credibility."[5] For many years, the cost of the maintenance of the files had been born completely by Fuller. When Fuller came to SIU, the university began paying for much of its upkeep.[6]

However, over the years, due to various financial setbacks and shifting priorities, the Library had been gradually transferring more and more of the cost of maintaining Fuller's papers back to Bucky. More problematically, however, was how the collection was being treated. Beginning in 1965, Fuller became increasingly concerned about where the collection was being stored and who was being given access to it. Bucky had requested that the collection be given its own designated space and be highly restricted. Copies of some terse correspondence back and forth between Fuller, Fuller's office, and representatives for the Library, still in the archives of SIUC, illustrate this litany of un-ironed-out issues.[7]

Hence, in 1971, with all these frustrations ongoing, Fuller left SIUC and relocated his base of operations to SIUC's sister campus, SIUE, in Edwardsville, Illinois. About a year earlier, as it became apparent that Fuller was not going to stay at SIUC, John Rendleman (of SIUE) offered him a position and office space at the Edwardsville campus.[8]

The disclosure of Fuller's move took place in September 1971 at a small press conference held in Edwardsville and attended by Fuller and Rendleman.[9] Five days

SIUE President John Rendleman is partially obscured by a gesturing Bucky Fuller at the September 1971 press conference announcing Fuller's move to Edwardsville. (*Courtesy: "The Edwardsville Intelligencer"/Madison County Archival Library*)

before the press conference, the *Southern Illinoisan* reported that Fuller would be opening an office in Edwardsville and that his current office, located at 206 West College Street in Carbondale, had already let go of some of its nine-person staff.[10]

In February 1972, the SIU Board of Trustees formally approved Fuller's "reassignment" from Carbondale to Edwardsville.[11] A few months later, in July 1972, the *Southern Illinoisan* announced that Fuller's Carbondale office would be closing permanently. Fuller's long-time secretary Naomi Wallace (who was about to leave Fuller's employ to work for SIUC's new president) had the duty of informing the press of the office's complete shuttering. The headline for this particular article, "All ties severed," bespoke of finality.[12]

Not surprisingly, also in 1972, as part of severing his ties, Fuller wrote to Morris Library and fully withdrew the earlier deposit of his Chronofile. The Chronofile's removal was not disclosed by Morris Library until months after it occurred. Today, the Chronofile resides at Stanford University.[13]

Shockingly, in regard to his house in Carbondale, at the time of Fuller's transfer, there was some consideration of actually picking up and moving his Dome out of Carbondale and over to Edwardsville. Whether this was to be achieved by dismantling the Dome, picking it up, and moving it on a truck or, as the military has done with some domes, via airlift, is unclear. Regardless, eventually, cooler heads prevailed and the Dome stayed put; even after Fuller's office move, officially at least, the Bucky Dome in Carbondale would remain his and his wife's home.[14]

However, by this time, Bucky was spending precious little time in Southern Illinois, and neither was Anne. If Mrs. Fuller was not traveling with her husband, she was, more often than not, spending time in California with her daughter and grandchildren. For both Fullers, their time in Carbondale was limited now to rare weekends.[15] It is perfectly reasonable that, consciously or not, both Mr. and Mrs. Fuller had curtailed their time in the area due to some soured feelings they both had about SIUC.

Over in Edwardsville, Fuller established his office. However, the space originally promised to him proved unavailable when he and his lone assistant, Shirley Sharkey, arrived. Instead, they found themselves relegated to some leftover Army barracks owned by the university but not even on the main campus. This was a major downsize for the Fuller operation—from two floors and a basement in Carbondale to just three rooms in Edwardsville. Those considerations and Bucky's still active travel schedule meant that he seldom came to Edwardsville at all. At times, it made more sense for Sharkey to go to him, in St. Louis, to meet him between flights to obtain his signature on letters and other vital documents.[16]

Actually, long before Fuller transferred to Edwardsville, he had been involved with the SIUE campus. In June 1961, he participated in an all-day conference on the commuter future of the college in Edwardsville, IL. Organized by SIU Design Department chair Harold Cohen, the event was held in East St. Louis in a plastic airdome. Fuller was one of the discussion's thirteen participants (some appearing on film) who debated the best way for the growing Edwardsville campus to serve its constituents.[17]

Though Fuller's "official" time connected with the Edwardsville campus would prove to be short-lived, he did make at least one long-lasting contribution to the college via the architecture for its Center for Spirituality and Sustainability. First announced in 1969,

The Fuller dome at SIUE in Edwardsville near its completion. (*Courtesy: SIUE*)

Fuller's design for SIUE was a geodesic dome. The design and construction of the dome was overseen by the firm of Fuller and Sadao, an architectural partnership consisting of Bucky and the late Shoji Sadao.[18] Even for a Fuller dome, the Center's geodesic sphere is distinct. The Center itself describes it:

> The building stands with surveyor-certified mathematical exactitude. Symmetrically astride the true planet Earth's "90th" western meridian of longitude. Standing under the dome our planet's continents can be seen outlined against a transparent blue ocean, as they would be seen by X-rays if one descended by elevator from Edwardsville, IL to the center of the Earth, always keeping Edwardsville directly overhead. The center's dome is also a true planetarium for the northern hemisphere.[19]

Completed in 1971, this still-standing dome is the only dome Fuller ever designed specifically for spiritual purposes. Today, it has been recognized as a local historical landmark.[20]

Then, in 1972, almost one year to the day after his move from SIUC to SIUE, it was announced that the University Science Center in Philadelphia had hired Buckminster Fuller as its "World Fellow in Residence." Fuller's full-time move to Pennsylvania was imminent.[21]

In late 1972, Bucky and Anne Fuller left Southern Illinois. They moved to Philadelphia where they lived in an airy, but square, apartment, not far from the Delaware River.[22]

Before decamping for Pennsylvania, as with most homeowners, the Fullers put their home up for sale. Hence, in October 1972, in the *Southern Illinoisan*, the following classified ad appeared:

> SPACIOUS HOUSING—For two in Geodesic dome home designed by Buckminster Fuller. Two baths, hot water heating in floor, corner lot, garden with fountain located near shops and University. The price has been reduced to $29,000![23]

This identical ad would reappear in *Southern Illinoisan* in November 1972 and again in April and May 1973.[24]

While the Fullers, then residing in Pennsylvania, waited for a buyer, they welcomed a short-term tenant. For a brief time, a research associate of Bucky's, the late Brendan O'Regan, lived in the Dome, though the main purpose of his residency seemed to have been packing up some of the Fullers' belongings and sending them east.[25]

Notably, as soon as it became evident that the Fullers would be leaving, there were those in town who immediately suggested that the home be taken over by the city or the university.[26] Fuller was already an icon and many believed that his house should be preserved in tribute to him. However, this feeling towards the home was not widespread enough at the time to garner the necessary support it needed.

Despite Fuller's celebrity—or perhaps because of it—not to mention the enviable arrangement that Fuller had with the university (his ample travel schedule, his paid-for support staff and other perks), there existed throughout Fuller's time with SIUC a certain degree of resentment against him by many others at the university. Often on the road or even overseas, Fuller was free from the day-to-day drudgery (and many of the responsibilities) of his fellow academics. In 1964, even *Time* magazine called

Fuller's duties at SIUC "vague and undemanding."[27] Yet, Fuller still got the majority of the attention from the press and the public, often overshadowing other members of the faculty.

The controversial nature of Fuller's role at the SIUC would linger long after his departure from the school and even after his death. Is it notable that, even today, with the exception of the one remaining picnic dome by Campus Lake, there is no geodesic dome anywhere on the SIUC campus and that nowhere at SIUC is there even much mention of Fuller's ten-year association with the university? Certainly, it seems there has been a sometime silence about Fuller and his one-time presence at SIU and in Carbondale. In 1995, the *Chicago Tribune* hit the nail on the head when they published an article titled "Odd Man Out: Buckminster Fuller Is Revered Around The World—Except In The Town He Called Home."[28]

Yet, since his exit from Southern Illinois, *c.* 1972, the legacy and personage of R. Buckminster Fuller has only grown. The issues that Fuller cared about most—like sustainability—have only gained traction.

Furthermore, thanks to Bucky's name, star power, and influence, an impressive coterie of highly influential names made their way to Carbondale, specifically to its Department of Design. These included radical architect Paolo Soleri and such forward-thinking psychologists as B. F. Skinner, Charles Ferster, and Israel Goldiamond.[29]

What should also not be negated was the extraordinary spotlight Fuller helped shed on this then-small Midwestern school during his tenure. From his *Time* magazine cover story to all the news coverage of his inventive local domicile to his plans to cover Manhattan in a giant glass dome-shell, every mention of Buckminster Fuller, in print or on TV, always described him as "Professor of Design, Southern Illinois University in Carbondale."

In response to his presence or not, the enrollment of Southern Illinois University more than quadrupled during his time with the college.[30] Without question, there was a sheen that Bucky brought to SIUC, and to Carbondale, that has not been diminished, and it is a patina far bigger than any dome Bucky Fuller ever lived in, built, or even envisioned.

The Bucky Dome in Carbondale would be on the market for almost a year. Then, in September 1973, Michael ("Mickey") Mitchell, with a loan from a bank in West Frankfort, co-signed for by his mother and father, purchased the home for $22,000.[31] Mitchell recalls it snowed the day he moved into his new home. Mitchell would own the home for the next twenty-seven years.[32]

According to Mitchell, the house was nearly completely bare when he assumed its ownership; Bucky left behind only his teapot and the home's refrigerator. Outside, the Fullers left behind their lawnmower and their outdoor tea table.[33]

However, new Dome owner Mickey Mitchell was just not any new homeowner with a taste for a good and interesting bargain. He says today, "Bucky and I were great friends. I bought the Dome from [him and Anne]. I promised them I would take care of it as best I could and preserve it."[34]

Mitchell first met Bucky around 1966 when Fuller took notice of the young man who so often showed up at his lectures, always taking a seat in the front row. Later, Mitchell became a close colleague, and for sixteen years, he was an (unpaid) travel assistant for Bucky. During his years accompanying Bucky on his trips, the duo would travel to Maine, California, Colorado, and Texas as well as, once, to London. Among

Above left: Bucky Fuller.

Above right: The front door of the Dome, not long after Mike Mitchell moved in. (*Courtesy:* M. Mitchell)

Mitchell's duties were audiotaping many of Fuller's lectures and musings (transcripts of which would later serve as the foundation for several of Bucky's books) and drawing many of the early visual elements for Bucky's World Game project. Like Bucky, Mitchell, a then twenty-eight-year-old native of West Frankfort, had forgone so-called "higher education" and, instead, educated himself, first, by reviewing the teachings and writings of such innovative thinkers as Bucky and Henry Ford and then by traversing the globe.[35]

Along with his interest in Bucky, Mitchell was also a musician. At the time of his purchase of the Dome, Mitchell was fronting his own music group, the World Man Band. Along with making good music, the band also hoped to foster good will, open thinking and many other tenets of the Fuller philosophy. Mitchell said in 1974, "Rock 'n' roll music is a way to design the environment. It's independent of nations—it's really an international language." He stated further that the band's purpose was to "put out the information in the environment that no one else is giving: world unity, how to fall in love with technology."[36]

To that end, Mitchell turned his bandmates, John Christy and Tony Huston (the latter the son of movie director John Huston), into roommates so they could work away day and night. Christy recalls:

My sleeping room was Bucky's upstairs library area. It was great! I was sleeping where Bucky spent a lot of time thinking and reading. Part of the main floor living room area was set up as a small four-track music recording area. It was also great.... [I]t was an inspirational space with uniquely wonderful acoustics.[37]

When they moved into the Dome, World Man was working on an album.[38] One of its songs, "Cultivate the Positive, Eliminate the Negative," was co-written by Mitchell's friend and mentor, Bucky Fuller.[39]

During his early residency in the Dome, Mitchell not only carried on many of Fuller's personal objectives, he even echoed the great man's design aesthetic. Mitchell decorated the home in a style he deemed "Dali Lama," an Eastern-inspired mix complete with tapestries and strands of Chinese twine hanging over the doorway going upstairs.[40] Mitchell's radical message and radical house got him profiled in the *Southern Illinoisan* in June 1974.[41]

During his two-decade plus ownership of the Dome, Mitchell estimates he put as much as $100,000 into the home's upkeep. One of his first tasks—sad but necessary— was to remove the house's ten overhead skylights. Even after all of Bucky's tweaks, the skylights still tended to leak during rainy weather.[42]

Yet even then and even after the installation of shingles, leakage remained a concern. Not only did the dampness prove an irritant but, eventually, it proved to be rotting to much of the home's wooden frame. Of particular concern were the tops (the "canopies") of the double patio doors around the home. According to Mitchell, they often demanded the application of silver tar—not black tar as that would crack in the sun—over their upper edges. Often Mitchell would instruct his later renters to purchase and apply new coats of silver tar and then have them deduct the cost of it from their next rental check.[43]

Still, water was persistent. According to Mitchell, he added a second layer of shingles—on top of the first—during his ownership of the home. The second layer of the shingles altered the exterior look of the home once more. White-grey shingles got covered up by brown-colored ones.[44]

If water was not a risk during his residency, then, at one time, fire was. Mitchell recalls, "The fan in the ceiling caught fire once at the top of the roof. If I had not been in the dome, it would have burnt to the ground." The small blaze resulted in Mitchell having to discard the home's original industrial fan and replace it with another.[45]

Other tasks Mitchell would take on in the Dome included spraying for bugs, replacing the boiler system, painting the exterior and the interior several times, and redoing the fence "a hundred times."[46]

The upkeep did not end there, however. Mitchell tried to keep the yard's small fountain operational as well:

When I moved in, I painted it blue. I replaced the water pump in the water spout. Bucky took pains to make certain the water was level and the water came out in an exact circle. It would make a perfect sound on the water as it hit back in the fountain and it was very relaxing.[47]

Mitchell adds about the house's outdoor amenities:

There was a driveway lamp that set to the right of the steps. It had a round hat on top so at night you could only see the steps and your night vision was not taken away, this was Bucky's sailor thinking to not go blind with a bright light.[48]

Mike Mitchell's friend and bandmate, John Christy, pictured in the Dome. (*Courtesy: M. Mitchell*)

Family and friends of Mike Mitchell enjoy the outdoors at a gathering at the Dome. (*Courtesy: M. Mitchell*)

The Dome's front-yard fountain as photographed by Michael Mitchell during his ownership. (*Courtesy: M. Mitchell*)

Mitchell lived in the Dome, off and on, for several years. Despite some of the extra work involved, he has no regrets about it:

> [Living in the Dome] feels like you live in Bucky Fuller's skull from the inside…. The dome was the helm of Space Ship Earth that Bucky ran the world from, drinking tea and writing like mad at his little table while Anne clanked the pots and pans in the background.[49]

A close friend of Mitchell's, Dr. Bruce Hector (who later founded the Carbondale Clinic), though he never lived in the Dome, visited it often. Dr. Hector says, "Actually, the first time I was in the dome was when Bucky lived there. I had attended many of Bucky's lectures. I was a student of Bucky."[50]

Though Hector had his own "bachelor pad" just north of Carbondale, hanging around in the Dome was more fun. Hector says, "I loved the open concept of it. All of my homes since then have been 'open concept.' I'm thinking about adding a domed deck to my current home here in California."[51]

Yet even at that stage in the Dome's lifespan, Hector still knew of some of the building's bigger issues. "Some of the plywood was already showing its age. The home leaked in spots. I think Mickey got more than he bargained for when be bought it," he says.[52]

The Dome.
(*Courtesy: M. Mitchell*)

Mitchell was able to get some assistance with the house around 1978 when he applied for and got a grant from the city for upgrades and repairs to the Dome. As explained by former city administrator Jeff Doherty, the grant program, an extension of HUD, was the Community Development Block Grant. It was gifted monies—between $7,000 and $8,000—presented to local homes for medium to low incomes; it was a means for cities to retain housing stock in their communities.[53] Among the requirements for the grant money was that the home had to be then owner-occupied and that all funds had to be used to fix code violations along with other alterations.[54]

Eventually, like the Dome's first owner, Mitchell began spending so much time on the road and not in the home, he decided to rent the place out and use its monthly proceeds to fund his travels.[55] It is believed that the very first renter of the Bucky Dome (and the third person in its history to live there) was Michelle Bach and her young family.

A dancer, choreographer, and teacher (much like Fuller's daughter, Allegra), Bach was born in Chicago. Her father was an engineer, so, growing up, Bach knew of Buckminster Fuller. After earning her first degree at another university, Bach relocated to Carbondale to further her dance education. Originally, she lived in Makanda, IL, along with her husband and young son. Yet, later, Bach began to think about moving into town:

I was good friends with the Roans, Herb and Sally, and their two daughters. A wonderful family. They were also good friends with Bucky and Anne. The Roans were the ones that told me about the Dome. They said, "You would be perfect for this!" And I was like "YEAH!"[56]

Bach recalls moving into the Dome in 1976; Mickey Mitchell, of course, was her landlord. Bach's Dome years were productive and inventive ones. She says, "The campus was so alive at that time. And the Dome was a very creative space. I created some of my best dances and theater pieces there."[57]

Since the Dome was still relatively new at the time she moved in, some of the home repair problems that would affect later residents were not an issue for Bach and family. Though the areas around the sliding glass doors were beginning to show their age, and Bach remembers at least one leaky pipe in one of the bathrooms, the Dome had few other issues.[58]

Nevertheless, Bach recalls that the red fence "swayed" some but was still standing and the home's cork floors were, then, fully intact. Bach loved the floor especially since the cork provided a good, somewhat buoyant surface to dance on. Plus, she adds, "It felt good to wake up and walk on that heated floor." The home's expansive floor plan also gave her ample room to create. Bach and brood even had space to move a grand piano into the living room.[59]

Bach used the loft area as her bedroom and gave the first-floor master's suite to her son. Still, she says, "The acoustics in that house were so good. No matter where I was in the house, I could hear him breathing just like he was next to me. It was like he was breathing in my ear."[60]

As much as she and her family enjoyed the inside of the house, they also enjoyed the closed-in yard. Bach says:

The Dome in winter. (*Courtesy: M. Mitchell*)

It was a beautiful backyard. I hung up a hammock out to the left of the fountain. The fountain was working at that time. Of course, to my son, the fountain was more like a swimming pool! Mrs. Fuller had planted the garden so you had flowers the whole year.[61]

Once, Bach opened the backyard up to visitors when she created and staged a site-specific dance performance that took place on the roof of the Dome. A solo piece, the dance featured Bach adorned in Christmas lights and, she says, "Was sort of parkour before people were doing parkour."[62]

Though the people that night were invited to the Dome, sometimes Bucky disciples showed up unannounced. Bach recalls, "Some people were nice, some were not…. I'd find them peaking in the windows, looking in at me."[63]

Along with her studies and dance work, Bach was also a businesswoman while living in the Dome. Bach's Gallery, a seller of limited-edition art and lithographs, was partially run out of the geodesic in August 1976.[64] Of her Dome life, Bach says:

I loved it! We had great parties at the house. I loved living in Carbondale. I even liked the weather! I remember lying out in the backyard in February! I wouldn't mind living in another dome someday.[65]

Bach and her family left the Dome in 1977 when she received a fellowship from the University of Wisconsin. Today, Bach (now Michelle Bach-Coulibaly) resides in Rhode Island, where, until recently, she continued to teach, dance, and choreograph at Brown University.

After Bach's departure, Mitchell was in need of a new tenant. Hence, in the *Southern Illinoisan* newspaper in May 1978, the following ad appeared: "CARBONDALE. Buckminster Fuller's Geodesic Dome, very private. Enclosed by high redwood fence. 2 Bedrooms & 2 baths. 932-3411."[66]

The ad would be repeated in a November edition of the paper and a few other times after that.[67] Journalist Audrey Block would, after seeing the advertisement, interview Mickey and write a story on the availability of the eco- and energy-efficient rental for the *Southern Illinoisan*. In her article, she recapped the history of the Dome and noted that its monthly rent would be $275. She also noted that renters, per the landlord, were free to redecorate if they wished. Block also mentioned that Mitchell was, currently, also attempting to entice SIUC into purchasing the Dome.[68]

Not long after the article, Mickey Mitchell decamped fulltime to the West Coast. Among the few possessions he took with him was the tea table he had inherited from the Fullers. One of Mitchell's first homes out west was, not surprisingly, a geodesic dome that he helped build on Saddle Peak Mountain. Later, he lived in famous music producer Lou Adler's guest house. Originally, Mitchell had moved to California to pursue his music, but with time, he gradually segued into such other endeavors as producing early concerts by Guns N' Roses and then, with Adler, running LA's Roxy Theater. Later, he founded his own tourist shuttle service which he runs to this day.[69]

Just before he departed however, Mitchell found his next renter. Based upon various records from the era, the late Linda Burzynski resided in the Dome for a time in 1978–1979.[70] A nurse by profession, Burzynski passed away in 2006.[71] While Linda was residing in the Dome, one of dome science's biggest supporters was also around, having made a hegira to the Dome shortly after it became a rental property.

Builder Ed Cook was living just north of Chicago in the early 1970s when he got bit hard by the Bucky bug. He recalls:

> I got a copy of the *Domebook 2* from a friend and I became *obsessed*. I went out and built my own dome. After it was built, I lifted it onto a panel truck and hauled it to Carbondale. I set it up south of Crab Orchard Lake and then I enrolled in SIU's Design Department.[72]

Unfortunately, for Cook, the year was 1974. "I found out after I got there," he says, "that Bucky had already left!"[73]

Still, despite that miscalculation, Cook's love for domes did not abate. As he studied building and design at SIU—and made friends with fellow student Thad Heckman—Cook continued to read up on dome technology. Soon, Cook was working for a local contractor, and one day, the office manager asked him a made-to-order question: "Do you know anything about these domes?"[74] The next thing Cook knew, he was being dispatched to "the shrine"—Carbondale's Bucky Dome. Cook's task was to work on the in-floor heating system and to repair the electric boiler.

Later, in 1976, Cook was again summoned to the Dome. That time, when Cook arrived at the house, he found another worker was already on-site and had "already taken apart a whole part of the dome—it could have collapsed!" Ostensibly brought in to deal with a water leak, this other handyman, instead, got carried away and began pulling down panels and then chords that could have undermined the central tension of the building. By the time Cook got there, the damage had already been done.[75]

Wanting to make everything right, Cook reached out to Geodesic, Inc., the then progeny of the former Pease Woodworking Company, the original manufacturer of the Dome. Cook said: "They still had all the parts and specifics. I made an offer to Mike Mitchell that for $10 an hour, I could copy the panels and replace them."[76]

Sadly, though, even $10 an hour was too much for the recently moved Mitchell. Instead, Mitchell had to turn to others. In fact, he turned to two unskilled, unlicensed "carpenters" whose work was sufficient but far from perfect.[77]

Harold ("H. B.") Koplowitz, a SIU grad, was another of the Bucky Dome's earliest renters. He lived in the Dome from September 1979 until November 1983 while holding down one of his first post-college jobs, that of feature writer at the *Southern Illinoisan*. It was in the Dome that Koplowitz wrote his notorious, highly enjoyable book *Carbondale After Dark,* a collection of essays and short stories about the town and its political and protest history.[78]

After a long career in journalism, today, Koplowitz lives in Florida, writing e-books and freelancing. The name of his company pays homage to his origins: Dome Publications.[79] A Carbondale native, Koplowitz first visited the Bucky Dome as part of a school field trip when he was in third grade:

> It was still being built, and my most vivid memory of the day is of the teacher warning us not to touch a mysterious substance on the walls that looked like cotton candy and seemed to be called "viper-glass." I learned later that was fiberglass.[80]

Writer and one-time Dome renter H. B. Koplowitz used the Dome's bedroom for sleeping and as a home office. (*Courtesy: Deb Browne*)

Since Dome owner Mickey Mitchell was usually in California, Koplowitz usually dealt with Mickey's mother (who would come over from her home in West Frankfort) as his primary landlord:

> She wrote the lease on notebook paper with a pen or pencil, and it was only a sentence or two long. The lease said the tenant was responsible for maintaining the fence and grounds. I didn't think anything of it until a storm blew through and took down some trees and half the fence. It would have cost thousands of dollars to fix it right. Instead, I got a neighborhood kid to help me (actually, I helped him) repair the fence as best we could, saving as much of the original wood as we could.[81]

As a Dome resident, Koplowitz found that "living in the dome was fun, often exhilarating. Even transcendental. Though many of his ideas seem complex or spacey, Fuller was at heart a pragmatist."[82]

Still, as in Bucky's day, dome living presented some challenges. Koplowitz once wrote, "The 1,400 square foot dwelling is modest, with but two closets, one bedroom, no basement and no garage … [and] it had a tendency to leak. With its odd angles, arranging furniture was a challenge."[83] Koplowitz continues:

> Many of the dome's features, like the air shaft and patio doors, were meant to utilize natural forces like sunlight and wind to control the temperature. The problem was that

sometimes they just didn't work. Even with fans, it was difficult to coax a gentle breeze through the house....[84]

After his first period of residency in the Dome—only a few months—Koplowitz briefly moved out. He states:

The next tenant stayed just a few months before going to Florida to join a religious cult. Thus, when I returned to Carbondale, in the summer of 1979 to become a reporter at the local daily newspaper, "Buckminster Fuller's historic dome" turned up in the "For Rent" section of the classified ads. That September, a co-worker and I moved in. He got the back bedroom and I took the loft.[85]

Unfortunately, as the weather grew cold, Dome living got even more complicated. The leaks that were a minor annoyance in the summer were a huge issue in the winter. Koplowitz says he always felt "exposed" in the winter months. He adds, "Any hope I had that the big patio doors would contribute significant amounts of passive solar energy in winter were dashed when I discovered icicles on them—on the inside!"[86]

The dog days of summer though were a different story:

Of all the seasons, the dome was least equipped to handle summer. Without vents in the top for hot air to escape (nor the bathrooms, for that matter), hot weather turned it into a sauna. The dark asbestos shingles soaked up sun all day and radiated heat into the night.[87]

Still, Koplowitz admits:

Nevertheless, dome dwelling was like living inside a bubble instead of a box. One felt just the right balance of being exposed to nature and protected from it at the same time. Besides, the bragging rights alone were worth a certain amount of inconvenience.[88]

During some of his residency in the Dome, Koplowitz shared his space with his friend Gary Marx. Marx was a SIUC Journalism School grad and, like H. B., was employed by the *Southern Illinoisan*. He lived in the Dome from 1979 to 1980.[89]

When Koplowitz approached him, Marx jumped at the chance to live in this "iconic" piece of real estate. He recalls, "I paid an ungodly cheap amount of rent—like $150 a month or something. H. B. took the choice spot of the loft for his bedroom but later regretted it because it got hot in the summer and he didn't have easy access to the bathroom."[90]

Despite having to do his part to keep the often-problematic Bucky fence upright, Marx enjoyed his time in the land of Bucky. He says, "We had a couple of parties there—nothing too wild and crazy. Living there certainly raised our cool factor."[91]

A few months after Marx departed, Koplowitz began sharing his home with his then girlfriend Deborah Browne, another SIUC student and the one who did the layout for the first edition of Koplowitz's *Carbondale After Dark*. She says, "I did it the old-fashioned way, on a light table with a T-square ... I imagined the vibes from the dome were helping me on this colossal project."[92]

The home at the time of Deb Browne's occupancy. (*Courtesy: Deb Browne*)

Browne, who remembers a monthly rent of $300, has vivid memories of dome-living: "The Dome had wonderful acoustics. If one wanted to hear what was going on in the front room, or when you were in the loft above, all could be heard."[93] She continues:

> I liked how there was a hanging pan rack also in a semi-circle opposite the kitchen cabinets. [Bucky's] wife also was responsible so they say for a wonderful garden, some of which was still blooming when we lived there, and a small circular, of course, cement pool for a fountain.
>
> When my twin brothers came to visit in '82 they immediately climbed up on the roof—*sans* ladder—and did a handstand. The way the patio doors were framed and shingled allowed for good handholds.
>
> I will admit it was odd for me to live in the Dome with its cave-like shape. I think if I had owned it and lived there I would do more with lighting the ceiling and using mobiles or something to enhance the space. Gazing up at the pentagons and hexagons was fascinating though.[94]

Koplowitz and Browne lived in the Dome until November 1983 when a $100-a-month increase in their rent ended their stay.[95]

As they did when the Fuller's lived there, the Poppen kids still lived next door to the Dome. They remember a steady churn of renters in the Dome over the years as well as those times that the home sat vacant. When empty, the Dome became a favorite, if

The Dome's interior during Deb Browne's time as a renter. Note signs of water staining visible on the ceiling. (*Courtesy: Deb Browne*)

illicit, hangout for some of the neighborhood kids. Lisa Poppen remembers, "We would sneak into the fenced yard and swing on the left-behind hammock and play around the fountain."[96]

Lisa's sibling Dylan concurs: "People would either leave the gate open or we'd skid under it. We used to look for four-leaf clovers in the yard. It was the nexus for about fifteen of us local kids. Sealed behind its fence, it was 'our' *Secret Garden* in its way."[97]

During one of the Dome's periodic vacant phases, it was opened up by owner Mickey Mitchell when a Japanese TV crew came to town to do a television program on Buckminster Fuller's life and work. Since Mitchell was in California then, he turned to his friend Bill Perk to help him coordinate with the videographers.[98]

In 1980, as his former home in Carbondale awaited a new resident, Buckminster Fuller, living in Philadelphia since 1973, was winding up his time with that city's Science Center.[99] During the nine years since leaving SIUC, Bucky, along with his work at the Center, continued to lecture and continued his far-reaching travel schedule.[100]

Despite his somewhat unceremonial exit from SIU, Fuller seemed to hold no hard feelings towards the university. He would return often in later years to speak on campus—usually to standing room only crowds. His first return engagement was in May 1973.[101] This was followed by other visits to Carbondale in 1976 and 1980.[102]

However, by 1980, Fuller was eighty-five years old and even his seemingly boundless energy was not immune from the normal aging process.

Finally, in order to be closer to their daughter, Allegra, who had long been a professor of dance at the University of California, the Fullers relocated to Pacific Palisades, California. There, Bucky adopted—for him—a slightly less hectic pace. He curtailed his work schedule even more when Anne's health began to fail. Twice in the early 1980s, Mrs. Fuller underwent operations for cancer. She was in the hospital and had just slipped into a coma on the morning of July 1, 1983, when Bucky came to visit her. Bucky was at her bedside when he suffered a massive heart attack and died. At the time of his passing, Buckminster Fuller was eighty-seven years old. Thirty-six hours after her husband's passing, Anne Fuller, too, passed away. The couple had been married for sixty-six years.[103]

The Fullers are buried at the Mount Auburn Cemetery in Cambridge, Massachusetts. They share a grave marker. On it, next to Anne's name, embossed in the marble is a single rose; next to Bucky's name, a circle divided into geodesic shapes is chiseled into the stone. Placed just above the main tombstone is a separate stone on which is inscribed: "CALL ME TRIMTAB"—BUCKY.[104]

Back in Carbondale, just days after the Japanese videographers left, a new tenant for the Dome was found. If city records from the era are correct, the Dome's next resident was the late Dr. Donald S. Wham and his wife, Elizabeth. They lived in the Dome from 1984 to 1985.[105] Dr. Wham was a native of Centralia and was employed by SIUC at its student health service from 1974 to 1992.[106] He passed away in 2003; Elizabeth died in 2011.[107]

Paul J. Hansel lived in the Dome next, from 1985 to 1986.[108] Though his bachelor's degree was earned at another university, Dr. Hansel got his master's degree at Southern and had just completed his Ph.D. in business (at the University of Kentucky) when he came back to Carbondale to teach in SIUC's Department of Business.[109] It was during his work on his MA, between 1977 and 1978, that Hensel first met Fuller during one of Fuller's later visits to campus:

> I met him at some school function or other. I very much knew who he was. I had started out thinking I was going to be an architect before I realized that wasn't where I belonged! ... We talked for a little while—maybe fifteen minutes.[110]

Years after, when he returned to SIUC to teach, Hensel conceived of living in Bucky's old home which, to his surprise, proved available:

> I honestly don't remember how I got in touch with the landlord, but, somehow, I did. We eventually worked out an agreement where I would move in and fix up a few things. The house had been empty for a while before I moved in.[111]

During Hansel's one-year occupancy, the unmarried professor painted the interior walls off-white and tried to get the water-pump circulator replaced. Hansel would also struggle with regulating the temperature in the home. As he recalls, "the heat worked a little too well sometimes."[112]

As would become increasingly common over the years, while living in the Dome, Hansel would often receive uninvited guests. "Every couple of weeks someone would

Above: Bucky and Anne Fuller—at the time of their deaths, they had been married for over sixty years.

Left: One of the few surviving photos of the Dome's original driveway gate; it was still operational at the time of Mike Mitchell's purchase of the home. (*Courtesy: M. Mitchell*)

stop by either because they knew of Bucky or just to see a geodesic dome home. They'd always ask to come in and I'd say 'Sure.'"[113] Hansel eventually departed the Dome when he got married. He says, "My new wife was really not interested in living in the Dome."[114]

After Hansel's departure, the Dome became home to University student William ("Bill") Clay. Clay grew up in Kentucky, Virginia, and then Chicago, where his father was a professor of philosophy at the University of Chicago. Clay was well aware of Fuller as his father was a futurist with a great interest in Bucky, Tesla, and other radical thinkers.[115]

The self-professed son of "hippies," Clay was something of a wild child himself when he first moved to Carbondale in the mid-1980s. A musician and fan of the Punk and New Wave scene, Clay majored in art before segueing into graphic arts and then into mass communications. The one constant throughout his SIUC career was his work at WIDB, SIUC's alternative radio station.[116]

When he first arrived in town, Clay and his then girlfriend lived in a non-dome home, but he always knew of the Dome and coveted it. Clay eventually made the acquaintance of then renter Paul Hansel. When Hansel let Bill know he was moving, Clay went home and told his girlfriend, "Honey, we're moving to Bucky's house!"[117]

After signing a lease with homeowner Mitchell, mailed to and from California, the couple moved in. However, living in "Bucky's house," came with a steep price tag. The rent was, as Clay recalls, a hefty $700 a month. Clay was a student and though he then worked two part-time jobs (one at local Mexican restaurant Tres Hombres), he still needed more money.[118] Hence, Clay took extreme action—he began selling marijuana.[119] Supplied with weed from growers down state, Clay found an eager clientele in Carbondale. Clay says today, "I really wanted to live in that house and that's the only way I could afford to do it." He adds, "I was just this anti-Reagan youth of the era and I'm not ashamed of what I did. I didn't hurt anybody. It's part of my past, my history."[120]

With his new influx of capital, Clay dreamed of using his proceeds to bring the Bucky house back to its original glory. He says today, "I wanted to re-open the skylights and get that gross stuff off the roof."[121] Yet even Clay's efforts weren't enough: "I tried to repair the front gate as best I could. I repaired the garden shed. There were vines growing and taking over the fence. The roof leaked."[122]

Clay was equally frustrated by what he saw as the city and the university's lackluster concern for the property. While living there, he tried to get the town to recognize the home as a landmark, all to no avail. He says, "It was like nobody even knew who Bucky was!"[123]

In fact, Clay's advocacy (and at-home "business") might have had the opposite effect on the then powers-that-be in Carbondale.[124] Rather than rally around him, the city seemed to have come down hard on him and his house on Forest Avenue. For example, twice in September 1986, Clay was cited by the city—once for having an unlicensed vehicle parked on the premises; the second for failure to cut the grass.[125] Clay says, "I think the City really wanted to tear the place down but I was a proud anti-establishment Gen X'er at the time and I wasn't going to let that happen!"[126]

As Clay also spent many of his college years working in the local music scene, he also recalls inviting such then-just-starting-out groups as REM, Red Hot Chili Peppers, and

Fishbone over to the house after they played their local gigs. Once, with Fishbone, Clay even staged his own benefit concert for the house. Unfortunately, not too much money got generated and, "The concert kind of got out of hand."[127]

Believe it or not, Clay even claims to have once seen the ghost of Bucky and Anne at the house. It was not long after he moved into the home:

> There were a group of us there—it wasn't just me—and my girlfriend and I looked out and, right there, by the fountain, were these two bluish-green figures. A man and a woman. He looked like Bucky and she looked like his wife![128]

Clay continues:

> It went on for a full minute. I had not been drinking either; I had a test the next day. Really, I was kind of scared. I looked at my girlfriend and she looked at me. I remember she grabbed my hand and I grabbed hers. We couldn't believe what we were seeing. I kept wanting to ask them, "Are you here? Are you happy?"[129]

Clay's college friend, Christian Blatter, may never have seen any ghosts at the home, but he visited it often. Blatter was a DJ during his college years, like Clay, and was involved in the local New Wave, Goth, and Punk scene. Blatter remembers the Dome home often rocking during parties as well as renter Clay often complaining about some of the house's features, like regulating its temperature and the non-working outside fountain.[130]

Actually, even before being a party guest, Blatter had a history with the Dome. Originally from Herrin, Blatter's family eventually moved to Carbondale. One day, back in the 1960s, Blatter's father was walking down Forest Street and Bucky himself was out doing some yard work. The two men struck up a quick conversation with Fuller, eventually inviting Mr. Blatter into his famous home. The elder Blatter found the famous Mr. Fuller open, interesting, and very talkative; they chatted for several hours.[131]

Bill Clay eventually got busted by the local authorities and departed SIU and the Dome. He became a paratrooper in the U.S. Army for a time ("It was a good thing: I need the structure and discipline," he says) and returned to military service in the wake of 9/11. Today, he lives in South Carolina and runs his own niche record shop and indulges his artistic side by buying and refurbishing antiques.[132]

Not long after Clay's exit, Mike Mitchell, in 1987, after fifteen years of owning the Dome, tested the waters of selling it—for a good price and an even better purpose. In May 1987, in the *Southern Illinoisan*, the following ad appeared under "Houses for Sale:" "BUCKMINSTER FULLER'S GEODESIC DOME located at 407 South Forest. Could be used as a museum or research center. Price reduced to $99,900."[133]

Mitchell also approached various architectural aficionados and Bucky devotees. Alas, there were no takers. Along with a periodic rental ad, the "For Sale" ad appeared in the paper for a few more months. In March 1988, it was noted: "Price reduced to $81,900."[134] Yet, still, this lack of a purchaser turned out to be good news for schoolteacher Linda Buchholz. She moved in in 1988—and stayed for the next seven years. Buchholz does not remember what she paid in rent but a classified ad from 1988 listed the monthly rate as $360.[135]

The Dome from the street.
(*Courtesy: M. Mitchell*)

A native of nearby Murphysboro who has a degree in elementary education from SIUC, Buchholz knew Bucky slightly; she was friends with some of the people in his office who, at the time, were working with Fuller on his EXPO '67 dome.[136]

Buchholz says she was well aware of the Dome in Carbondale and had always wanted to live there.

> I was always checking to see if the dome was available. I was living in a small apartment and had had enough of that. Then, one day, I was coming down Forest and happened to see the "For Rent" sign![137]

Buchholz quickly called Mickey Mitchell. Mitchell referred her to his then rental agent, located in an office in downtown Carbondale. The agent gave Linda a house key and sent her, alone, to the then empty Dome. Soon after, she signed a one-year lease with Mitchell and renewed it verbally every year thereafter.[138] Even today, over twenty-five years after her days in the Dome, Buchholz, now living on the east coast, remains excited about the house. She says, "I loved living in the dome! I liked sleeping up in the loft. The whole home was so peaceful."[139]

Still, dome-living presented a few challenges. Buchholz discovered that most of the home's driveway gate ("what remained of it") was off its hinges and was lying to one side. As was also quickly becoming the norm, the famous Fuller fence also often presented issues. Buchholz says, "It was mostly standing but was still something of a hazard."[140]

Some other areas that demanded attention during her years included a leak in one of the bathrooms; the need to replace a pane in one of the sliding doors; and the regular spraying for wood ants, who were, she says, "devouring" parts of the building.[141]

Buchholz attempted any necessary repairs, deducting the cost of her supplies from the monthly rental check she mailed to Mitchell. Once, after doing some mending on one of the interior walls, Buchholz returned home later in the day to find a note from her landlord pinned to the front door. It read "Good job!" That was the closest she came to ever actually meeting her landlord.[142] As Mitchell was residing full time in California during much of his ownership of the house, few of his renters ever met him.

Oddly, it was not anything with the actual Dome that proved to be the biggest irritant for Buchholz while she lived there. Instead, it was various reactions to her living there. Buchholz recounts people being jealous of her tenancy, even going so far as to harass her to leave:

> One night, a friend stopped by and, after he came in, he said, "Are you expecting someone else?" I wasn't and when I looked outside there was just this man standing right there in the dark, in the driveway, just outside my door. When I came out, he went away.[143]

Some people, she says, just seemed to resent the presence of the Dome and "that big red fence," as Buchholz describes it, sitting in the neighborhood. Over her years there, Buchholz encountered various students throwing things at the house among other acts of mischief. She says, "I think some people just really had a problem with this unique, strange home in a neighborhood of otherwise traditional homes."[144]

Part of the attention, good and bad, that Buchholz and her home was on the receiving end of may have been due to a sudden amount of attention Buckminster Fuller started getting around 1989. In May of that year, SIUC's Campus Museum announced a large-scale, multi-month exhibition on Buckminster Fuller. The exhibit *Ideas and Integrities: R. Buckminster Fuller at Southern Illinois University* opened in February 1990 and featured, against a black, yellow, and chrome color scheme, original Fuller sketches, maps, poetry, photos, blueprints, and models.[145]

Buchholz only eventually moved out of the home when the upmost tip, the "polar-pent," of the home, began to leak and water dribbled down into the living room. "I knew," she says, "that fixing that was beyond my abilities." So, reluctantly, after a tenure in the Dome second in duration only to Mr. and Mrs. Fuller's, Buchholz departed from the home in 1995.[146]

While, for many years, for Mike Mitchell, the Dome did provide a (usually) steady stream of income, he always saw his role not so much as landlord but as caretaker. He adds, "I always did my best to get [renters] that would take care of the Dome. They had to promise to take care of it." Mitchell estimates that, during his renting of the Dome, he had only about ten renters, all paying between $275 to, eventually, $700 a month.[147]

Though, over the years, Mitchell encountered a few people interested in the house or the land it was on, he always refused to sell. Mitchell's retention of the Carbondale home was his way of keeping Bucky's legacy alive. He always hoped that the home would be purchased by a larger organization and repurposed into a museum, or other tribute, devoted to Bucky. As noted, long before he offered the home for rent, he first offered it for sale to SIUC, but the university declined. At one point, Mitchell also began to solicit for a corporate benefactor who would buy the home, donate it to the city or the National Trust. Those attempts, however, also went nowhere.[148] The building of the necessary appreciation for the Dome would take time.

Still, by the mid-1990s, Mitchell had grown weary of owning the Dome. The tipping point for him arrived when the home failed an inspection by the city of Carbondale. The city ordered that the home could not be rented out again until certain additional repairs were made. Mitchell decided he just did not want the long-distance hassle anymore.[149]

Additionally, and more importantly, the upcoming year—1995—would mark Fuller's centennial. Interest in Fuller would certainly be at its height and the possibility of getting an appropriate buyer for this major piece of Bucky memorabilia seemed feasible. So, in mid-1994 (and again in 1995), Carbondale's Bucky Dome was placed on the market for $100,000.[150]

Mitchell's original listing agent for the home was Cheryl Gilberg. An SIUC graduate with a degree in architecture, Gilberg was excited to represent the house. She says, "I saw Bucky speak on campus when I was a freshman!" After Mitchell mailed Gilberg a key, she visited the Dome for the first time.[151] She recalls, "It was a very cool house but it was not in great shape at that time. And I did think the price he was asking for it was too high."[152]

Mitchell's starting price might have seemed steep for a one-bedroom house, but, in true home-selling tradition, he never expected the full amount. He did, however, hope to recoup some of the money he had put into the Dome over the years.[153]

Before any sale, though, Mitchell first hoped to have the residence designated as a local historical landmark by the city of Carbondale. Achieving such status would mean that the home could not be destroyed or altered, and that the integrity of the original structure would be protected by law. Obtaining local landmark status would also pave the way for the next levels of recognition—state and national landmark status.[154]

Obtaining such recognition, from any entity, is a process of application and review. Acting locally, Mitchell filed, on behalf of the Bucky Dome. His first application for the honor with the city was in the summer of 1995.[155]

Also pushing for the home to be recognized by the city was Bill Perk, a retired SIUC professor of community development and of design, and a longtime Bucky-phile. Harry F. W. "Bill" Perk had first learned of Buckminster Fuller from growing up and reading architectural magazines:

> Then I heard Bucky lecture in Royce Hall at UCLA in the Fall of 1958, and that changed my life thereafter.
>
> Such an extraordinary occasion!
>
> I made it my business to keep up with him after that and it was Fuller, as a matter of fact, that helped bring me to SIUC.[156]

Perk was born in Los Angeles and served in the military in Korea. In early 1947, back stateside, he enrolled in UCLA, where he earned a degree in physics and chemistry in 1951. Later, at USC, he got a master's degree in city and regional planning. Back at UCLA, he began work towards a Ph.D. in Urban Land Economics.[157]

From 1956 to 1959, Perk was employed with the Rand Corporation in LA. Later, he joined the planning department of the Bank of America. Later still, he joined the organization the Council for Better Building. It was there, one day, that he fielded a call from Allegra Fuller Snyder, daughter of Buckminster Fuller. Bucky was giving a lecture at Beverly Hills High School for members of Bennington College's alumni. Would the Council share their mailing list to spread the word of her father's talk? Perk obliged and Bucky was forever grateful to him. A friendship ensued.[158]

In 1960, Perk went to work for Lockheed Electronics and relocated himself and his family to New Jersey. About four years after that, Perk found himself "at liberty." His friend, Bucky, said, "Why don't you come to SIU?"

Perk replied, "But I don't have a Ph.D."

Bucky said, "Designers don't need Ph.D.s!"[159]

Perk joined SIUC's Design Department in the fall of 1964 and he would remain for the next thirty-one years. For a time, he even served as the Department's chair.[160] Perk states, "I visited Bucky and Anne in the dome on several occasions, one of which was when he asked me to read the proof of his about-to-be-published book, *Intuition*, which I did while sitting at his dinner table in the dome."[161]

After the Fullers left in the early 1970s, Perk hoped that the home would be reestablished either as a museum, research facility, or, perhaps, even an inn. At the very least, perhaps, the home could obtain a full-time caretaker.[162]

Over the years, Perk, who had long been acquaintances with Mickey Mitchell, would occasionally be alerted to the Dome's rental availability by Mitchell, as Mickey pursued new tenants.[163] Nevertheless, Perk was "surprised" in 1995 when he was approached by Mitchell about finding possible buyers for the home.[164]

Perk, unfortunately, knew of no "wealthy angel" to come in and buy the house. Instead, Perk quickly mobilized a different sort of effort. Via his own network of friends and a message he put in the local newspaper, he recruited Bucky friends and fans as well as others concerned with the long-term well-being of the Dome. Together, they hoped to find another solution.[165]

Early meetings of the group—many featuring "people I didn't know," according to Perk—were held at the Dome and then regularly every month for the next six months.[166] Like Mitchell, Perk agreed that one of the first steps in realizing the Dome's endurance was to have it labeled "historic" by the city of Carbondale.[167]

Linda Gladson was a senior planner for the city of Carbondale from 1992 to 1997. Today, Gladson lives in St. Louis but still owns a home in Carbondale. A graduate of SIUC's Art and Design program, Gladson was a fan of Bucky's and, today, owns a dog named in his honor—Buckingham Fuller ("Because there's only one Buckminster Fuller," she explains). Gladson, then with the city, was in favor of designating the Bucky Dome a historical landmark, an action that was possible in that Carbondale had just recently become a "Certified City" within the state.[168]

Yet despite considerable support (Gladson says, "No one in their right mind wouldn't have supported that decision"), the proclamation did not go through.[169] At the very last moment, just hours before the Council meeting, homeowner Mitchell sent a letter to the Council withdrawing his application. Having read in the press about some of SIUC's recent expenditures, which he viewed as wasteful, Mitchell became incensed at the university's reluctance not to take a greater interest in Buckminster Fuller or the Dome. He stated in his letter: "The dome is not receiving the respect and attention that it deserves, and it would be hypocritical for the dome to be placed on the registry right now."[170]

Another view of the Dome.
(*Courtesy: M. Mitchell*)

Along with being disappointed in what he viewed as the community's under appreciation of Fuller's importance, Mitchell also halted the landmark application because obtaining it would impose further (often costly) restrictions on him as an owner, impacting the repairs and methods he could carry out on his own property: "It would have left me in a bad place. It just gave the local area control of [the property], via many codes and rules, that would have been difficult to deal with from California."[171]

After his withdrawal, Mitchell nevertheless became resolved to protect the Dome as best he could. He soon implemented the necessary repairs needed for the Dome to be rentable again and continued to pursue an appropriate new owner for the dwelling. "My problem was how to get as much of my investment back as I could," he explains, "and make sure that I kept my promise to Bucky and Anne to do my best with their dome."[172]

Despite the kerfuffle over the Bucky Dome as a local historic site, the 100th anniversary of the birth of Buckminster Fuller did not go unobserved in Carbondale. Local Bucky fan Cornelius Crane, who at the time described himself as a "graduate student from Earth," spearheaded many of the activities that celebrated Bucky in the town Fuller once called home. During the week of Fuller's 100th birthday, a banquet was held in Bucky's honor. Also held was a sunset concert and a special playing of Fuller's World Game. Carbondale's Mayor, Neil Dillard, also proclaimed July 12, 1995, as "R. Buckminster Fuller Day."[173] Perk and Crane both stated that they hoped the festivities surrounding Bucky's upcoming birthday would become an annual event.[174]

The various centennial celebrations occurred with great success and the controversy over the home's "official" status as a landmark died down. Meanwhile, interest in finding new ownership of, and the overall state of, Carbondale's Bucky Dome continued. Unfortunately, though, things did not always progress smoothly.

Though, at that time, Bill Perk was not necessarily looking for a home to buy, he was still the one leading the local charge to purchase the Dome. Hence, in August 1996, Perk and his newly configured group of (approximately) twenty-four people began to work to either buy the Dome themselves, or find someone with deep enough pockets who would.[175]

When an article about the new Dome group appeared in the local paper, the paper referred to the group as "Friends of the Fuller Dome"— it was a name that stuck, even being reprinted on the group's first "official" stationery.[176]

As the new group got its footing, an early piece of good news fortified them: the Carbondale Park District informed them that if an individual could be persuaded to purchase the Dome, and gift it to the city, they would take on the responsibly of maintaining it.[177]

Some of the Friends' first efforts ranged from trying to drum up interest locally to attracting national and even international interest, preferably from one or two benevolent millionaires. The group also examined their own bank accounts to see if pooling their own resources would allow for a purchase. Unfortunately, according to founding Friend Cornelius Crane, many of the early "friends" were young people or even SIUC students—not groups known for having money reserves.

Bill Perk, meanwhile, hoped to renegotiate the price of the home. At one of the group's meetings, a "sympathetic" realtor in attendance finally said to Perk, "Make Mitchell an offer to move it forward." Newly inspired, Perk made his way to Carbondale's City Hall and looked up the 1972 price that Mitchell paid for the house. Perk remembers:

> Since Mitchell was then seeking $100K, I took 22/35th of that and proposed $68K. The realtor in our group said we must first see if the bank would loan us that amount, so I asked my bank who said no loan could be made without an appraisal. So I paid for an appraisal.[178]

That appraiser was Barbara Zieba, an independent appraiser then living in nearby DeSoto. According to Zieba, she had a "difficult time" putting a value on such an unusual home. Not only was the home not in the best of conditions, there was little else to compare it to. Zieba finally had to compare it to area A-frame homes and to other homes in the region of "similar age, square footage and location." Eventually, Zieba had to reach out and learn the value of dome homes in other states.[179]

Ultimately, Zieba valued the house at considerably less than Mitchell's $100,000. She pinpointed its value at $46,000. Zieba told the local *Daily Egyptian* newspaper that the home's then status of not being on the historic registry undermined its overall value. Mitchell, however, in that same article stated, "Whether it is 'registered' or not, it is still a historic house."[180]

Regardless, though, the gap between the appraised value and the amount the Friends had acquired was too gaping. Though Carbondale's First National Bank promised to finance 85 percent of the appraised $46,000, they would not consider financing a loan above $68,000. So, says Perk, "everything came to a halt."[181]

For the next five years, the Bucky Dome remained in the hands of Mickey Mitchell. However, misdelivered mail between the city of Carbondale and Mitchell in California (who for a time lived on a houseboat) caused a variety of Dome-related issues to fall between the cracks.[182] In February 1995, the city, in its latest inspection, listed a litany of items about the house that needed immediate attention:

> outside light fixture broken
> tub faucets leaking and loose tile around [tub]
> no vent fan in bathroom
> screens (windows & doors)
> indoor structure failure (above sliding doors)
> bannister loose going upstairs[183]

These citations were not excessive or exclusive to the Dome. Due to its status as a college town and the large section of its population that are students, the city has strict rental inspections, and like dozens of other homes within the city limits, the Dome sometimes got caught up in policy.

John Yow was one of those city inspectors during this time. Though he visited hundreds of homes during his long career, he remembers the Dome not only due to its famous provenance but also since, for a time, the city's then-mayor, Helen Westberg, lived just down the street from it. Yow was often unnerved by some aspects of the house's condition. "It sometimes felt neglected," he says, "I always wished we could get someone interested in it."[184]

Matters related to the Dome were not helped in the court of public opinion when a 1995 *Chicago Tribune* article relayed an unsubstantiated report that the city of Carbondale was considering bulldozing the house.[185] Despite the bad publicity, Mitchell saw to the latest repairs to the home and, once again, started renting out the Bucky Dome, mainly to students.

Two of those renters were then-married students Pete DePaoli and Kim Alexander. Before they arrived in Carbondale to go to school, the Chicago-raised Kim came to town in the spring of 1995 to house and apartment hunt. Though not familiar at all with Fuller, she nevertheless "got excited" when she came across the rental ad for the geodesic dome in the local paper. She was able to secure the "really cool" home by calling up Mickey Mitchell, and not only offering to rent the house, but to pay six months of rent in advance.[186]

Pete and Kim moved into the Dome on June 15, 1995. Later, the couple worked out an arrangement with Mitchell to stop paying rent and, instead, "pay" via the work they did on the Dome.[187]

The DePaolis were thrilled to be living in a piece of history. Before they moved in, the couple was given *carte blanche* by Mitchell to assist with the home's upkeep. In their efforts, the DePaolis even went so far as to post on a Bucky internet fan page that they would gladly welcome some help with a few small repairs:

> Greetings to all! My husband and myself (Pete & Kim) have just signed a contract to lease Fuller's Dome in Carbondale IL. The dome is requiring some rehab to pass city inspection and we are inviting any interested parties to assist in that rehab. The pay

Kim Alexander, a one-time renter of the Dome, poses in front of it. Note that in the back of this photo, the corner wooden shed, added by the Fullers to shelter a lawnmower and garden tools, is faintly visible. (*Courtesy: K. Alexander*)

would be the satisfaction of having been involved in the refurbishing of Fuller's dome that he built and lived in for some years while a professor at SIU. The first weekend of the work will be May the 5th. Please write us back at althea58@aol.com. Thanx and I love this list! P.S. When the dome is finished and we are moved in (June 15) we will have a dome party to celebrate and keep the dream alive…. The city inspector said if it had been let go for 5 more years, it would have been the BULLDOZED!!! GADS!!!

We cannot let that happen!![188]

Among the items needing the most attention, according to Kim, were some inside leaking issues, a few bad chords overhead, a disabled sliding door, and the outside fence.[189]

Later, Kim DePaoli announced on that same Bucky website that once she and her husband got more settled, they would host an open house in order to share the experience of the Dome with others. Kim said at the time, "We will make the Dome available for lots of visiting in a controlled manner. [Otherwise,] it's like owning the Mona Lisa and hiding it in the basement."[190]

During their time in the Dome, the couple also encountered the occasional unannounced visitor who knocked on the front door and wanted to see inside "Bucky's house." One time, they were approached by local TV Channel 8 News who wanted to do a profile of the dome-dwellers.[191]

While in residence, the DePaolis repainted the inside of the Dome—replicating it white—and painted the outside flat sides of home "forest green." Kim says, "The blue wasn't making it anymore."[192]

Pete and Kim would live in the Dome until late 1996 when they relocated to Cobden, IL. Kim, an SIUC grad with a degree in University Studies, now lives in Murphysboro and stills gets passionate about her two years on Forest Avenue. She says, "I loved living there! I liked living in a geodesic dome. I wish I had another one."[193]

Not everyone who aided in the Dome's upkeep, however, actually lived there. In an arrangement that both men now label as odd, Mickey Mitchell, for a time, had a caretaker for the house who paid him.

It was 1995. The aforementioned Cornelius Crane was an SIUC grad with a degree in automotive technologies, graduated in 1994. During most of his time in Carbondale, he did not know of Buckminster Fuller. Later, however, when working a part time job planting trees for the city, a co-worker asked him if he knew the name Buckminster Fuller, because "You sound like him in many ways."[194]

Crane was intrigued and began his research into the work of Fuller. Later, when a stranger on the street saw him reading Fuller's *Critical Path* ("For the second time," Crane adds), the man said, "Well, if you're interested in Bucky Fuller, you should go talk to Bill Perk."[195] Crane did.

With the photographer standing in the loft, Kim and her dog pose on the first floor of the Dome. (*Courtesy: K. Alexander*)

For Crane, meeting Perk was like meeting a kindred spirit. Together, they discussed Fuller's philosophy and lamented the fact that no place within the city did Carbondale pay tribute to one of its most famous ex-citizens.[196] As Crane recalls, for about eight months in 1995, he acted as the Dome's "official" caretaker, paying Mickey. All the time, Crane says, he was just wishing he could purchase the house outright.[197] In the meantime, the Bucky Dome went back to being rented.

Colleen M. Doyle studied music education at SIUC and graduated in 1995. She came to live in the Dome when a guy she was dating spotted a "For Rent" sign on its fence. She recalls, "I think we were at a yard sale in the neighborhood when my then-boyfriend brought it to my attention. We talked to the guy living there and he got me in touch with Mike Mitchell."[198]

Doyle had driven by the home numerous times and was familiar with Buckminster Fuller. She was intrigued by the "novelty" of living in such a home. "I remember the first time I walked in, I thought, 'Where do you hang a picture?'"[199]

Talk of turning the home into something other than a rental space was already afoot during Doyle's time in the Dome: "I remember there were a couple of meetings at that time, at my house—I mean Bucky's house—with Bill Perk and Hugh Muldoon, talking about getting the home on the National Register."[200]

Slowly, the idea of preserving the Dome gained momentum. First, a pro-Bucky Dome editorial in the *Southern Illinoisan* called the home "an asset worth preserving." Then, for a short period of time, Carbondale's kid-oriented Science Center considered moving into the Dome after they lost their old location. Yet downgrading from their then existing space of 4,700 square feet to the Dome's approximate 1,400 square feet proved too big of a change.[201] Today, the Science Center resides in Carbondale's University Mall.

As with many of the residents who preceded and followed her, Doyle wanted to do what she could to maintain the Dome:

I really worked on it. Mainly on the outside. When I got there the driveway was completely overgrown with brush and weeds. I cleared it all out. It made a world of difference. The fence was starting to fall down and we patched it as best we could and replaced the slates like Bucky originally had them.[202]

Internally, Doyle repainted the interior floors—"a gun-metal grey"—and enjoyed the radiant heating in the foundation. "That feels *phenomenal* when you are walking around barefoot," she says. Though water stains were visible at some vertexes, Doyle maintains, "I never witnessed a leak."[203]

Doyle eventually left the Dome to take a job in Cobden. After some time there and out of state, Doyle, now Colleen Doyle-Parrott, is now back in the Carbondale area and teaching in local school district 95. She and her husband now keep up with local Dome news and even have a small collection of Bucky-related memorabilia.[204]

After Doyle exited, SIUC student Martin Moran moved into the Dome in his junior year of college (1997–1998). Studying Construction Technology, Moran was already familiar with Fuller. He says, "I lived originally over on Hays, so I was aware of the home. I walked by it often. Finally, one day, I went and knocked on its front door. The woman who answered said she was moving out at the end of the year and referred me to Mike [Mitchell]."[205]

Excited, Moran contacted Mitchell and quickly signed a lease. He paid half the rent which, for him, came to $350 a month. Moran says, "That was high rent back then but well worth it, being in Buckminster's home."[206]

"I had one roommate.... We flipped a coin and he ended up on the losing end of having the upstairs library as his room. I had Bucky's master suite on the first floor."[207] Moran had a great time during his one year of residency:

> It was a great hit with my clique of hippie friends. At the bottom of the stairs in the Dome was the couch, a recliner, the TV and, for us, a little basketball hoop.[208]
>
> Matt, my roommate, though, probably didn't care for that as he had no privacy up in his room and the noise from the TV and all.... He didn't know anything about Bucky, so I set about teaching him....[209]

One of Moran's favorite aspects of the home was how verdant its surroundings were. He says, "It was very shaded. The trees completely covered the drive. It was like this hidden part of a crazy campus."[210]

The trees around the home also seemed to help with the home's temperature regulation. "I'm sure I had an AC wall unit in the bedroom," Moran says, "[But] I don't remember any problem being cool or warm."[211]

The Bucky Dome—now painted green—during the time Marty Moran lived there. (*Courtesy: M. Moran*)

I had great intentions of further fixing up the dome when I was there, in particular the leaks, the dilapidated fence and the fountain in the backyard. But college life and no money got in the way. I loved living in the Dome. I wish I could have stayed longer but money and some other issues prevented it.[212]

Moran's roommate was Matt Staulcup. Staulcup was the one on the losing end of that coin toss. Moran and Staulcup had known each other since they were kids in the Chicago area. After rooming together on Hays, the two made a beeline for the Bucky Dome when it became available.[213]

Staulcup still gets excited over the fact that he once lived in Bucky's house. He says, "It was awesome!" Staulcup, now living in northern Indiana, got so "Fuller-fied" during his residency he even named his cat "Bucky." Staulcup says, "It was fabulous living there! We kept the house and yard up nice. The neighbors were so grateful for that, they used to bring us cookies."[214]

Shortly after the duo vacated the Dome, it became home to its next set of renters: Troy C. Smith and his college sweetheart, Alexis Holle. Smith and Holle occupied the Bucky Dome for a year, May 1998 to May 1999.

Smith, an SIUC accounting major, was not familiar with Buckminster Fuller but his girlfriend, Alexis, was. Holle, then a grad student in art and design at the university, had grown up in Springfield. She had long known of Bucky since her parents and some of her earlier art teachers were all fans. Not long after she arrived in town, she went looking for the famous dome home.[215]

Later, while others were renting it, Holle and Smith attended a few parties at the Dome. Soon after, Holle actively pursued the chance of living there herself. She went so far as to semi-frequently inquire with her campus acquaintance, and then Dome renter, Marty Moran, about if or when he was planning to move. Holle, now a wedding planner living in Arizona, says:

I really wanted to live there and I think I stalked them! But in those days, in Carbondale, good rental houses were hard to come by. You had to move quick.[216]

Then, one day, I happened to be reading the *Daily Egyptian* and I saw that the house was for rent! I immediately found a pay phone and had to use a calling card—what we had in those days—to call the landlord in California.[217]

Holle got ahold of Mitchell and, "for the most part, he leased it to me then and there."[218] Holle and Smith coordinated their moving dates with Moran and Staulcup and, despite, according to Smith, the "absurdly" high rent (around $600 a month), the young couple moved in.[219]

Holle recalls, "We knew that there was going to be some upkeep with the house and we thought we were handy at the time." Yet they soon changed their minds especially in terms of the fence which, when they arrived, "was standing—for the most part, though there was leaning here and there."[220]

Holle remembers once repainting the fence in a shade close to its original reddish-brown. The new paint job, however, was not enough to always appease the city. Though enough of the fence was standing to keep the couple's dog in the yard, the city did cite them a couple of times for the fence, which the city saw as a decaying eyesore. Smith simply remembers forwarding the notices they got onto Mitchell. Holle remembers Mickey as "always on top of" the house's repairs.[221]

Smith states, "We painted the entire inside and set up the loft for Alexis' art studio. She did some beautiful work there."[222] Some water staining on some of the inside walls necessitated the use of the landlord and renter's best friend—Kilz paint, a mixture that works well at covering even the most persistence discoloring. While the interior walls were painted white, Alexis went bold on the first level. With most of the home's original cork floor gone by this time, exposing the concrete underneath, Holle painted most of the first level floors (living room and kitchen) a bright red.[223]

The couple also undertook other improvements, once getting the home sprayed for termites. Smith and Holle, however, did nothing in relation to the fountain. Smith says, "If it rained, there'd be water in there and the birds would take a bath!"[224] Holle remembers later covering "the pit" (the fountain) with a piece of plywood so that their dog would be safe.[225]

Later, during the "sweltering" summer, Smith finally "called a buddy" and installed a window AC unit in the bedroom. It required a little cutting of the original window frame but was necessary to make the home livable.[226]

In a reoccurring theme for the Bucky Dome, water leaking into the house was sometimes a problem. Though Holle remembers very little seepage during her tenure, some leakage did occur over the south-facing sliding door, so Smith and Holle simply put all their potted plants under the jamb so that they would be naturally watered.[227]

At least twice, Smith recalls, during their time in the Dome, people from the Dome's history would stop by. He says, "Some random person would just knock on the front door and say, 'I used to hang out here with Bucky.' They'd always want to see the place again and we'd let them in."[228]

Along with putting a roof over their heads during their college years, Smith and Holle also have a special reason to reflect fondly on the Bucky Dome. Smith relates, "My eldest son was conceived there!"[229]

Though Smith and Holle have now gone their separate ways, they share not only a son but fond memories of the Bucky Dome. In a wonderful denouement, their son, Tony, now college age and weighing many career and life options, did, at one time, show an interest in architecture.[230]

Sometime around this time—*c.* 1998—William Denton moved into the Dome and resided there for "about three months." An SIUC student at the time, he walked by the "cool" house "hundreds" of time and, one day, after spotting a "For Rent" sign, he reached out to "this guy who lived on a boat in California."[231] That guy was, of course, Dome owner Mike Mitchell who rented the abode to Denton for $600 a month plus a promise by Denton to see to much of the maintenance and upkeep.[232]

Denton began—like most of the renters—with high hopes and great ambition. He painted some of the floors; he repaired the railing up in the loft; he did what he could with the fence out front. However, the variety and drastic state of some of the tasks soon got the better of him. Adding to his stress, Denton found himself somehow listed by the city as the property manager for the house. That meant that if the city found any violations, Denton was on the hook for them.[233] At first, Denton forwarded all the citations onto Mitchell out west, but Mitchell, often traveling, did not always get to them that fast but the city kept coming.

Eventually fed up with the leakage, the "moldy and nasty" kitchen sink, and the front fountain "that never worked," and tired of being harassed by the city, Denton departed the Dome.[234] Still, despite his short time in residence, William looks back with great fondness at the "great party house" he was briefly affiliated with and the luck he had in being so close to Bucky.[235] After a time residing in Florida, Denton has since returned to the Southern Illinois area and loves seeing and visiting the home today as it is being revived.[236]

The next renter of the Bucky Dome was Kevin J. Connor, a professional electrician, he lived there from 1999 to 2000. Connor grew up in northern Illinois and had lived in Florida before he journeyed to Southern Illinois. He was a latecomer to Bucky but eventually immersed himself completely. He says, "I've read every book Bucky ever wrote. I'm just sorry I never got to meet him."[237]

Living in a mobile home in Carbondale, Connor jumped at the chance to move into the Dome when he saw an ad for it in the *Daily Egyptian*. Connors reflects, "It was more than I was paying for rent then but I was making pretty good money at the time." Connor believes his rent was around $400 a month.[238]

Connor later made the acquaintance of Holle and Smith before they moved out; like them, he did his best to keep the place up. He repaired the lights that lit the driveway, tried to get the fountain going again (but, sadly, did not succeed), and did the best he could with the fence.[239] Inside, he repainted the walls white, and since he worked at the local Lowe's hardware superstore, he used his personal know-how and employee discount to tape and mud some of the joints that needed filling.[240]

Connor's fondest memory of the Dome however was the time that his young granddaughter visited him, and she discovered she could run the entire circumference of the house. "That was the best! The day she was running and discovered that!" Connor says.[241] Connor says today, "I'm so glad about the Dome being restored! I was afraid we were close to losing it, it would have been like losing Thoreau's cabin in the woods...."[242]

Connor would prove to be the Bucky Dome's penultimate renter and the last under Mitchell's ownership. Mitchell looks back today and is grateful to all of them. He said recently, "I was very lucky. I really had a good bunch of renters over the years."[243]

Is it significant that the Dome has attracted so many creative types? That not only did Bucky enjoy some of his most productive years there, but that Mickey Mitchell composed music there, Michelle Bach fashioned dances, H. B. Koplowitz wrote, and painter Alexis Holle enjoyed one of her own creative peaks while living inside it? Former NFP president Cornelius Crane, for one, believes that there is a connection:

> Fuller talks about [how] human beings radiate an ultra-high frequency wave that comes out of us. This dome is just soaked with Buckminster Fuller's energy, and I think when people come in here they feel it themselves.[244]

Former NFP president Brent Ritzel concurs:

> I've had people tell me they've stayed in domes for a while and they feel a sort of higher state of consciousness. Bucky was king of touching on that; he tried to keep everything grounded and technical, yet he didn't reject the metaphysical.[245]

As noted, as tenants came and went from it over the years, the Dome underwent various changes. The original fence extension that separated some of the driveway from the northwestern corner of the Dome eventually disappeared. Similarly, at some point, Bucky's original storage shed (triangular in shape and nestled into that same northwest corner of the property) collapsed and was carted away (its "footings," though, can still be seen in the outside concrete).

As mentioned, the outside color of the house also changed. Originally two tones of blue with a white roof, the roof morphed in shade with the various layers of shingles. Then, in the mid-1990s, the light blue external walls were repainted a "forest green." Even the front fountain changed color with various paint jobs; first, it was blue, then white and then aqua.

The interior colors of the Dome have also fluctuated. Originally white, it is believed to have stayed that color during all of the Fullers' residency. However, sometime before H. B. Koplowitz moved in, the inside walls were altered to pale blue. Photos from Koplowtiz's era in the home bear this out. Later, the walls would be returned to white or off-white.

The floors of the home also endured some changes. In the beginning, pressed cork squares covered all the floors including those in the upper loft area. However, at some point, after the departure of the Fullers, the cork in the kitchen and living room vanished. Stories vary. Some state that a renter's dog chewed up most of the first-level flooring. Another story goes that water staining made some of the first floor's cork unsightly. Finally, there is one story that states that various "accidents" from a resident's two Dobermans caused the floor to become damaged beyond repair.[246] In any event, today, only the upstairs study's cork remains fully original and intact, and though painted cork remains intact in the bedroom, only a small portion of the original cork has survived completely untouched; it is inside of Bucky and Anne's bedroom closet.

Without its cork covering, the exposed concrete floor on the first level of the house welcomed various paint jobs as well—in time, it traversed from concrete white to gray, and finally to red.

For the Dome, finally, in 1999, after over twenty years of ownership, and after a few false starts, Mickey Mitchell became convinced that it was time for him to let it go. He recounts, "My parents had both died…. I decided it was time to cut the cord from Carbondale and it would be better for everyone and everything."[247]

Though Mitchell briefly entertained two offers to buy the home—one from California and one from an architect in Kentucky—what he had long hoped was that the home would actually go to fellow Bucky-phile Bill Perk: "I knew Bill could get the historical and city backing needed thanks to his university connections."[248]

Once again, at this time, Perk, aided by Cornelius Crane, attempted to interest SIUC in purchasing the property.[249] Perk has said, "It ended up the university was unable to deal with it even if I gave it to them, they didn't have the funds to deal with it, and even as a gift it didn't look like that was an option."[250] Perk once again rallied fellow Friends of the Dome, and also again, negotiations to purchase the home from Mitchell began in earnest.[251] Yet, once more, terms were not easy to come to.

Though everyone agreed that the home was invaluable, how to be fair to all and get the home the protection that it needed left the parties at a deep impasse. Mitchell still believed the home was worth $100,000. Meanwhile, Perk knew that neither he nor any of his comrades could come up with that amount.[252] Finally, after much back-and-forth, both parties agreed to a $50,000 sale price. Perk says:

> Having retired from SIU and federally required to "disperse" a certain fraction of my retirement funds each year, I realized I could borrow from my broker enough to enable me to buy the dome from Mitchell for his $50K and be able to pay that loan back in three years. No bank required. Thus, I became the new owner of Bucky's dome in 1999--with a tenant on a year's lease.[253]

The sale of the Bucky Dome from Mickey Mitchell to Bill Perk was finalized in July 1999.[254] The sale came complete with the home's original abstract, signed by both Anne and Bucky, which Mitchell gifted to Perk at the time of the transaction.[255]

Of course, the house also came with a tenant—Kevin Connor had just moved in; Mitchell had leased the home one last time (in June 1999) just before the sale. New Dome owner Perk allowed Connor to stay until the end of his lease, plus some additional time on a month-to-month basis. This arrangement worked out well for both men—Perk could generate some income from the home and Connor could keep a roof over his head.[256] Kevin Connor finally moved out of the Dome in 2000 when he decided to relocate to Florida to be closer to his family. Before leaving town, he handed the keys over to Bill Perk.[257]

Remarkably, in this age of house-flipping and short-sales, during its lifespan, the Bucky Dome—now well over fifty years old—has had, basically, only three owners: Bucky and Anne Fuller, Mickey Mitchell and, finally, Bill Perk and the "Friends of" not-for-profit.

Looking back, ideally, of course, as soon as Bucky and Anne departed Carbondale, their home should have been taken over by someone or something able to see to its long-term care, or, failing that, the home should have been quickly purchased by a professional architect, social scientist, contractor, and carpenter (and, preferably, one who also happened to have very deep pockets), who also loved to spend every single weekend doing fence repairs. Unfortunately, that did not happen and history does not work like that. Ideal outcomes seldom occur, and it always takes time for appreciation and perspective to take hold.

In that regard, Mickey Mitchell (too often ascribed as an "absentee landlord" in press about the Dome) did his duty and achieved his goal of "taking care of the Dome" for the Fullers and for posterity. It was not easy, and while, as a rental property, the home may have endured greater wear and tear due to its turnover in tenants, Mitchell's long-term custodial hold of it kept the home safe from others who might have had little regard for Bucky and what Bucky's abode symbolized.

Mitchell said recently, echoing the lessons of Fuller, "Do your own thinking to help others, helping yourself is death."[258]

3

RENO- (1999–2019)

reno-: ren·o·vate (ˈrenəˌvāt) v. restore (something old, especially a building) to a good state of repair; refresh; reinvigorate.

When the Bucky Dome was sold in 1999, it was sold in an "as is" condition, which, despite the best efforts of Mitchell *et al.*, was, according to Bill Perk, "not very good by then, and only got worse after the last tenant moved out."[1]

Founding NFP President Cornelius Crane added, "It's probably a credit to the Dome that it's still standing at all."[2]

Ironically, there is some (sound) speculation that Bucky Fuller was fully expecting to raze and replace his Dome about fifteen years after he built it. The Dome was, to him, an experiment.[3] Perk has stated, "Bucky would have torn it down and replaced it with his newest prototype. He thought of shelter systems as most people regard automobiles."[4] Fuller also, no doubt, believed that, in ten to twenty years, his own approach to engineering, as well as new building technologies, would have advanced enough to render the original home obsolete and ready for an update. For Bucky, everything was always a work in progress.

As the Dome's new owner, Bill Perk had many "first orders of business." After the paperwork was signed, there had to be a complete assessment of the structure. The dream was to bring the house back to the look (inside and out) it had when Bucky lived there, but first, what was the basic structural integrity of the home—walls, wiring, and pipes—now, forty years on? What had to be done to bring it back? How much time was needed? How much money would it cost?

Shortly after the Dome was acquired by Perk, Jon Davey (an SIUC Architecture professor and recently added "Friend of the Dome") described the home's situation to the *Daily Egyptian*: "The building is like the body after a bad car wreck, you don't try to preserve it or put it back together, but you work to stabilize it."[5]

The Dome, pictured in the *Southern Illinoisan*, around the time it was purchased by Bill Perk. (*Courtesy: Southern Illinoisan*)

Former SIUC Professor Bill Perk, the Dome's third owner.

As in most triage situations, time was of the essence.

Despite the presence of shingles (two, and, in some places, three layers thick) and other innumerable repairs to the roof, water leakage was still a Dome problem. Though not quite sieve-like, water trails and brown staining was visible at many of the internal vertexes and evidence that moisture had long been seeping in. Much of the home's first-floor cork was also pock-marked with water damage or was just missing completely. Wood rot was also extensive and even, in some places, visible to the naked eye. Water damage was especially extreme above the "canopies," the areas just above the sets of sliding glass doors around the home. There were even places where separation had taken place in the home, creating gaps. These gaps were most evident over the canopies; in fact, by positioning oneself in the right spot, one could see daylight through the roof of the south canopy.[6]

Additionally, from the outside, a quick scan of the roof showed various places where other patching had been done over the years—often with mixed results. Architect and later NFP board member Thad Heckman noted the following up on the roof:

> I was working on the roof and literally put my foot through a completely rotted-out area. Of course, I had to repair the damage but that was actually fortunate. It allowed me to see the complete history of roof treatments all the way down to the original roof deck. It was painted; brush strokes could be seen in the original weather coating material that looked like nothing more than a heavy "deck paint," a coating that one might use on, say, a wood plank front porch of that era. The coating didn't appear to be substantial enough for long term resistance to the elements at all—especially a geodesic dome with a hard southern exposure.[7]

Above left: After Bucky's original industrial fan in the home caught fire, Mike Mitchell replaced it with a more traditional ceiling fan as seen here. Note, in this photo, that water damage is visible on the ceiling.

Above right: As seen in this photo, eventually, the walls of the Dome began to separate from the Dome's roof. From outside the Dome, one could clearly see through the gap into the home's interior. (*Courtesy: Kim Alexander*)

One of the home's worst areas of leakage was at its very top apex or "polar-pent," as it has been labeled. There, the outward facing pentagon was, according to Heckman, "severely rotted—mushy, in fact."[8] That hub would eventually have to be completely replaced.

Ultimately, it was determined that the home contained seventeen different sources of major leaks. During the worst downpours, water collected from inside could be collected by the bucket.[9] Lest anyone believe though that the home's porous tendencies indicates poor design on Bucky's part, bear in mind that many of the buildings of Frank Lloyd Wright (including, ironically, his Falling Water) often leaked.[10] Also, the renowned I. M. Pei (he of the pyramid in front of the Louvre) once struggled with a skyscraper he designed; the giant windows of its upper floors kept falling out and raining glass onto the street below.[11] No designer is infallible.

Furthermore, Fuller was often at the forefront of new approaches in building, and at the time that his Carbondale home was built, some of the materials he used were not yet proven, especially for water protection.

Interestingly, when Heckman repaired one part of the roof, he found something surprising: "Some of the Celastic taped joints were still quite tight at the edges after all of these years—of course they had also been protected below the shingles for many of those years."[12]

Water and wood were not the only issues, however. Time and the elements had also done damage to the home's basic structure. As resilient as domes are, they are not unfailing; a compromising of one vertex can cause the entire form to become destabilized. Even to the untrained eye, the Dome, from some angles, appeared sprung. Over the years, many of the

The "polar-pent" (or upmost top of the Dome) eventually had to be completely rebuilt.

Dome's renters (often without the knowledge of Mike Mitchell) repeatedly took on some well-intentioned but still ill-advised repairs. Yet, these short-term fixes often resulted in doing long-term harm.[13] For example, over the years, some of the home's triangular panels had been removed, sloppily repaired, or even replaced with incorrect geometry.[14] Yet in a dome, any change in measurement can have a domino effect to its overall shape. The changes caused an effect on the home's overall curvature resulting in, as noted by *Architect Magazine*, the home having an "egg-shaped profile."[15]

All of these concerns were just in regard to the home's exterior. Inside, things were not much better. The walls were badly spotted with water stains throughout. The hood for the kitchen range had long ago been taken down (as it was not working) and was now sitting in the bedroom closet. Among other early discoveries, much of the home's original cork flooring had been worn down, destroyed, or completely removed. Only the cork in the home's upper study area, and a small portion in the bedroom, remained completely intact.[16]

The eventual removal of some damaged pieces of wall and floor, however, was not without some pleasant surprises. Behind one section of drywall in the bedroom, workers found a previously unknown of metal name plate nailed to a vertical stud; it marked the home as a genuine "Pease Dome." The plate will stay where it is and a glass pane over it will make it visible forevermore.[17]

Then, as if all that were not enough, the Dome continued to be the victim of vandalism. Located in a part of Carbondale that many used as a shortcut to the bars, the home's notoriety has often made it a target of mischief. In the years the "Friends" have owned the Dome, a brick has been lobbed through a glass door and there has been at least one break-in and burglary. At least once or twice, kids have scaled the roof of the

Above left: The kitchen as it appeared when the home was purchased by Bill Perk. In this photo, the oven's range hood is missing. (*Courtesy: E. Long*)

Above right: The guest bath of the Dome before its restoration. (*Courtesy: E. Long*)

Sample of some of the wood rot visible at this time.

Dome. As recently as 2014, a board member, found that a booze bottle had been lobbed in through one of the doors and, later, what was thought at the time to be a full can of beer—something seemingly unconscionable for a college town.[18]

In March 2001, the city got an anonymous tip that part of the redwood fence surrounding the Dome had been "knocked over" by the wind and was now "hanging" loosely. The city acted quickly since the home was now in violation of local ordinances. Dome owner Bill Perk was issued instructions to either fix the fence, have it removed, or be fined. In a communique to the city—composed on "official" "Friends" stationary—Perk thanked the city for alerting him to the problem and noted that the problem was "more likely a result of human activity, not 'wind damage.'" Perk promised to get right on the necessary repairs. Perk also went on to name the citation's author, neighborhood inspector Chuck Cremeens, an "Honorary Friend of Fuller's Dome."[19]

Then, just before SIUC's Spring Break 2001, Perk discovered someone had tried to break into the home by using a broom handle as a lever.[20]

Concerns like that and about the overall well-being of the home sitting empty caused Perk to borrow a page from previous owner Mitchell and find a renter for the home. That arrangement would generate money for the fledging foundation while also acting as an on-site defender of the Dome.

The previously unknown-of metal plate noting it as a Pease Dome; it was found behind the wall in the bedroom.

The (in)famous Bucky fence. (*Courtesy: White & Borgognoni*)

Justin R. Edgren has the distinction of being the last resident of Carbondale's Bucky Dome. An SIUC grad with a degree in political science, Edgren learned of the home's availability via a friend, one-time Carbondale City Councilwoman Maggie Flannigan. Then residing in Chicago, Edgren was looking for an out-of-town retreat for long weekends where he could write music.[21]

Edgren moved in in the summer of 2001. He paid $300 a month with the agreement that he would also do some house upkeep. Edgren got an idea of what would be involved with "upkeep" the first night he moved in. He says, "I was sitting on the couch that night and it started to rain outside. Suddenly it was like a waterfall down my back!"[22]

During his years in the Dome, Edgren repainted the floors (a maroon-ish-red shade, according to him), patched the roof with tar paper, and did "lots of caulking." The outside fence, though, was a lost cause. It was dotted with holes and some of its slats were already loosely stacked in the yard. The waist-high gate originally near the home's front door had long ago disappeared. Edgren says, "The house was ... barely inhabitable. It got pretty overwhelming."[23]

Today, Edgren, now a professor at Lindenwood University, admits that he was not overly familiar with Buckminster Fuller when he first moved in. However, that changed once he arrived—thanks to a steady stream of impromptu visitors:

> People were always coming to the door wanting to see the house. At least seven or eight people from all over the country showed up over the years. One time a student film crew from South Korea knocked on the door. One time, a person I remember referred to the house as the "Mecca of Architecture."[24]

Despite the home's state of disrepair and the interruptions, Edgren proved productive in the home. He said, "I got a lot of writing done; the home's acoustics were great. I did some work with sound artists like Josh Gumiela. The house had good vibes."[25]

Edgren would live in the Dome for about two years, but even though his presence warded off vandals, that was not the only threat against the Dome. Structural issues and the weather posed bigger problems. Due to these factors as well as time and money, Bill Perk, Cornelius Crane, and their associates realized that something aggressive had to be done to prevent further damage to the Dome. One day, the men mentioned their quandary to Thomas Zung, who offered an inventive solution.

A renowned architect, Zung is president of Buckminster Fuller, Sadao and Zung Architects and was a student of Bucky's. Some of Zung's buildings over the years have included the General Motors Headquarters and the John F. Kennedy Center for the Performing Arts.[26] To aid the preservation of the Dome, Zung suggested the novel idea of building a second dome to be placed over the existing Dome, giving the original home some all-over protection.[27]

Hence, in October 2001, Bill Perk submitted a zoning certificate to the city requesting permission to erect a "temporary construction shelter" on the Bucky Dome property. This shelter would be a new dome made to fit over the existing dome—the second dome being a just slightly larger half sphere. To explain the idea to the City, Perk enclosed with his application a simple, hand-drawn diagram—basically a circle with a larger circle around it—to show them what the board had in mind. After the over-dome's erection, there would remain, around the perimeter/walkway, a 4-foot gap between the two nesting structures.[28] Once installed, and covered with a giant tarp, the over-dome would protect the actual Bucky Dome from further weather erosion.

For the building of the "over-dome," Zung suggested the group turn to Blair Wolfram of Dome, Inc. Located in St. Paul, Minnesota, Dome, Inc. is one of the leading providers of dome structures throughout the world. Blair Wolfram is the company's co-founder and, perhaps, dome science's biggest booster since Bucky himself. Wolfram built his first dome in 1982 and he has not stopped since. His personal residence consists of nine domes including his house, garage, a boat house, and a jungle gym for his kids.[29]

Now hired for the project, Wolfram, utilizing the same geodesic principals as the original Bucky Dome, installed the steel skeleton that would go over the existing home. On the Forest Avenue site, he secured the outer dome to the ground with augured rods, fastened to the bottom edges of the steel shell with nuts and bolts, then overlaid the over-dome's rigging with chicken wire bent and molded with pliers to fit to the curvature underneath. Over that, yards and yards of greenhouse plastic was spread until full coverage was achieved. Besides being large enough, the greenhouse plastic was chosen for its surprising resiliency. Strong against rain, wind, and solar degradation, the special plastic is made to hold up for three years, but it actually proved even stronger than that; it would not need to be replaced for almost a decade.[30]

The building of the over-dome took place over SIUC's fall break of 2001. As in Bucky's day, this dome and its installation was efficient. Wolfram, who traveled in from Minnesota and helped set up the over-dome himself, remembers, "Everything was brought in on a trailer—the steel on one pallet, the plastic on a second pallet and the chicken wire, a third."[31]

Again, as in 1960, the putting up of the over-dome was a one-day affair. The final covering was achieved via the help of some volunteers that included Eric Smith, Steve Belletire, Jason Rangel, and Brian Bundren, among others. "It was like building a giant erector set," Belletire, an SIUC professor of industrial design, said at the time. Also just like that day in 1960, on site during the over-dome's build was original contractor/builder Ira Parrish whose surprise presence was a special treat for all.[32] Others there that day were group board member Larry Busch and the student news group Alt News 26.46 who filmed much of the day's happenings.[33]

However, not everything went smoothly. At one point, up on the Dome, Wolfram noticed a depression in a particular part of the roof. One of the building's "hexes" was dimpling inward and could collapse. Wolfram and his team had no choice but to take immediate action to prevent a cave in. They quickly installed a tall 6 × 6 wooden post inside the Dome, wedging it against the floor and the wall. It extended up to the ceiling where it was "blocked" (configured) to support the underside of the roof.[34]

Though the post ruined the interior aesthetic of the home, it did prevent the Dome from falling further inward. By sheer necessity, the 6 × 6 would remain in place for the next thirteen years. Later it was determined that the dent was caused by an ill-fitting repair undertaken previously and hastened by a poorly placed foot during the installation of the over-dome.[35]

The final full cost for the entire over-dome was about $5,000. Regardless, the protective dome went on to earn back its cost many times over. For example, only a few years after the over-dome was installed, one of the home's north side trees fell onto the house. It weighed so much that it actually bent one of the over-dome's metal chords.

Above: The over-dome, with its first plastic covering, *c.* 2003.

Left: The intrusive support beam inserted into the living room to keep the ceiling from falling in. (*Courtesy: White & Borgognoni*)

Had the protective shell not been there, the limb would have crushed part of the original Dome.[36] Along with justifying itself in time, the over-dome would also earn an off-color nickname—the "con-dome."[37]

Believe it or not, Justin Edgren was still living in the house when the over-dome went on. At first, despite his now-obscured view, Edgren did not mind the over-dome's presence as it discontinued overhead leakage and it helped regulate the temperature. However, with time, the outer shell affected airflow and that began to cause him problems. He says, "I think there must have been a lot of mold in there. I developed health issues. I had to have multiple air purifiers."[38]

Eventually, it became necessary for Edgren to move out. The board wanted to move ahead with its restoration and use the home more as part of their fundraising efforts. Edgren understood and left in the summer of 2003. As mentioned, Edgren was the last person who ever called the Dome "home."

As the Dome was no longer a rental, it was now, per city ordinance, free from city inspections. Still, the home would still be the topic of routine patrols by inspectors and to neighborhood alerts. In 2010, the property was twice cited. Once for a loose sewer cover which was fixed by the city, and then, when someone spray painted on the property's construction fence; this latter problem had to be fixed by the board.[39]

Once the Dome was covered and unoccupied, it allowed the Friends to refocus on the bigger issue: determining the various repairs that the Dome needed. One of the first people Bill Perk reached out to on this issue was Joe Clinton. Clinton was once a grad student at SIUC; in 1963, he was enrolled in the school's School of Engineering. During his time at the university in the 1960s, he worked on a contract from the National Aeronautics Association, and Bucky Fuller was one of that project's advisors.[40]

Coincidentally, Clinton would later depart from SIUC the same year Fuller did: 1970. Clinton moved to New Jersey and pursued his Ph.D. at NYU. For fourteen years, Clinton was the coordinator for the Department of Design Technology at Kean University. Clinton has also worked as the advisor to the trade organization, the Dome Home Manufacturing Association. Over the many years, Clinton remained ("off and on") in touch with Bill Perk.[41] With his great knowledge, "Bucky roots" and geodesic know-how, Bill knew he needed Clinton's advice. Clinton recounts:

> In the beginning—I think I was [at the house] for three to four days—I was to survey the damage. Just do a visual estimation of the problems—there were panels with dry rot--and then I looked over some of the existing repairs. I worked out some of the geometry that the house was built upon. I worked with Thomas Zung some, trying to see if the Pease people were still around.[42]

Despite his status as the home's owner, Bill Perk always saw his role with the home as intermediary. He was not interested in being the Dome's long-term owner. Yet that decision brought with it other questions: If not Bill, just who—if anyone—was going to live there? Who—or what—were they saving the house for? Who was going to pay for it?

Turning the Dome into a museum was one idea. In its just-purchased days, there was also a thought about turning the home into short-term housing for visiting dignitaries

Above: A shot from the kitchen looking into the bedroom. Note the presence of the air-conditioner installed by a renter; it has since been removed. (*Courtesy: White & Borgognoni*)

Left: The hole in the roof.

or university scholars. Maybe it could be a bed-and-breakfast. While contemplating these problems, the Friends began to explore individuals or entities to gift the Dome to.

Southern Illinois University, Carbondale's biggest employer as well as Bucky's one-time academic home, seemed like the perfect recipient for such a gift. When Thomas Zung visited SIUC in October 2001 to speak about his book, *Buckminster Fuller: Anthology for the New Millennium*, he implored the university to take a more active role in promoting Fuller's thinking and to preserve the home. Zung said, "There is a need for the public, especially the young, to discover his thinking anew."[43]

Yet despite respect for what the Dome represented, interest in taking over the home by the University, or anyone else, for that matter, was not easy to come by. The board's founding president, Cornelius Crane, along with Perk, made the first formal proposition to SIUC and the city for one of them to take over the Dome. Together, Crane and Perk met first with SIUC's then President John Guyon. But though there was great interest among various university "higher-ups," Crane and Perk knew that SIUC taking on the Dome was a long shot. Another historic home in the area (in Cobden) had also recently been offered to the university and been declined. With that precedent, Crane and Perk knew that the Dome becoming a Saluki property was unlikely.[44]

The men next met with then Carbondale Mayor Neil Dillard, but even though they got a meeting with him, they never got any follow-through from community officials. There had just been a semi-scandal about a proposed new city golf course, paid for by taxpayers; hence, the stage was not well set for the city taking over the Dome.[45]

Even a conversation with the SIU Foundation—though pleasant enough—proved, unsuccessful. According to Perk, the Foundation cautioned the new group about such a donation: should SIU's Foundation take over the Dome, they would then have total control over the property—even the right to tear it down if they wanted to. Certainly, this was not in the board's best interest in regard to preserving Bucky's legacy.[46]

Crane also discovered that other funding avenues were also fraught with problems. The organization was founded just after the terrorist attacks of 9/11. The subsequent financial uncertainly that gripped the nation afterward did not bode well for this upstart charity.[47]

Soon, finding that the group was in need of some more specialized advice, Bill Perk made a phone call to Mike Jackson, manager of the State of Illinois's Historic Preservation Agency. Jackson offered to come down to Carbondale and offer guidance to the group.[48]

At the meeting, Jackson commended the group on their intentions and noted that a historically recognized site can often mean as many as 25,000 annual visitors to a city. Yet, for all the good news, Jackson also had concerns. Among them, the Dome was in a residential neighborhood; if it was reinvented as an attraction, how would it deal with a large influx of people? How effectively could Bucky's philosophy be conveyed in such a, relatively, small space?[49]

If those issues were not daunting enough, then what of the restoration's price tag? At the time, it was estimated that a new dome could be built for around $30,000, but in order to restore the Bucky Dome, it would cost upwards of $250,000.[50] Until funding sources could be found, Bill Perk's retirement nest egg served as the primary source for funds.[51] Later, Crane underwrote many expenses via his own professional salary.[52]

Over the years, the group has approached or been approached by a variety of TV shows and magazines anxious to rehab the Dome as a TV special or a photo feature. However, often, these entities have had their eye more on speed than historical accuracy. Therefore, these offers have been declined.[53]

Still, despite the hiccups, 2002 was a good year for all things Bucky, at least in Southern Illinois. Along with developments regarding his local home, Carbondale Community Arts sponsored a local exhibit. *Thinking Out Loud: Artifactual Excerpts from R. Buckminster Fuller* was mounted at Carbondale's s University Mall and featured a scale model of Fuller's proposed St. Louis-based Old Man River City project as well as his screen prints, dating from 1927 and 1928, which reflected his interest in air routes.[54]

Meanwhile, the Friends of the Fuller Dome continued to meet regularly at SIUC's Morris Library, Carbondale's Public Library or Italian Village, the Carbondale pizzeria that would become a *de facto* headquarters for the group.

New recruits also came to the cause. Unlike many early board members, Joni Reeves was not an SIUC grad or even an Illinois native. Hailing from Ohio, Reeves got her undergraduate and advanced degrees in engineering via colleges in Ohio and the U.S. Air Force Institute. She learned of Bucky during her engineering studies.[55]

After living in Colorado for many years, in 2002, Reeves came to Carbondale and went to work for the University's athletic offices; that same year, she joined the board. Reeves would remain with both until 2008 when she returned to Colorado.[56]

When Reeves first arrived in Southern Illinois, her friend "Corny" Crane took her to the Dome. "He was very proud," she says. Reeves remembers being struck by how sad it was that the city was not doing more to see to its long-term care. "It's one of the few true treasures that that city has and to see how they'd allowed it to decay," she laments even today.[57]

Soon, Reeves was donating her time to the Dome's restoration. One of her early tasks was at the exhibit at the mall. Though Bill Perk was the exhibit's primary "docent," when he could not make it, Reeves often covered.[58] Once an "official" "Friend," Reeves found herself acting as the group's ad hoc secretary. Then, later, Reeves would have the good fortunate (?) to move nearby to the Dome and, hence, she often found herself checking in on things there. Reeves also became a one-person cleaning crew. She says, "I tried to get the kitchen a little nicer so we could have more events there." Often, she worked late into the night. She says, "It didn't creep me out too much, despite it being barren and damp."[59]

Reeves also took on many other jobs. On behalf of the group, she filed some of its first grant applications. Additionally, she says, "I joined before they [the Friends] were even a recognized organization and I tried to blaze a trail to get them to become a not-for-profit. It took a lot of spin sometimes." However, Reeves was heartened by those around her. "I was so impressed," she says today, "with the passion of the people who were working to improve—save—this landmark."[60]

Reeves' work paid off. The RBF Dome NFP—Not for Profit—was incorporated in July 2002.[61] Cornelius Crane, who had been involved with the Dome's preservation since 1995, drew up the first application for the Foundation, filing the necessary papers with the IRS. In his paperwork, Crane had to work to convince the Internal Revenue Service that this organization was fully on the up and up. With Bill Perk still the owner of the property, there was a fear that the foundation would be wrongly viewed as a tax

Several layers of shingles are visible in this shot of the Dome's roof. (*Courtesy: White & Borgognoni*)

A 2003 meeting of the "Friends of the Dome" inside the home's kitchen/dining room. *From left*: Bill Perk, founding "Friends" President Cornelius Crane, Joni Reeves, John Johnson, and Robert (Bob) Swenson.

Interior of the Dome's living room at the start of the restoration project.

shield, a personal project for Perk. As a result, Crane had to repeatedly make it clear that Perk was intending to donate the home to the NFP and that the board would be doing all the restoration themselves.[62]

Shortly after the NFP became official, Crane became the group's first president. The board's other members were Bill Perk, Joni Reeves, and John R. Johnson. The late John Johnson came to the board via his friendship with Bill. Perk recalls:

John Johnson was a grad student in the Design Department when I arrived in 1964. I soon realized he was smarter than me and definitely more accomplished as an actual designer! While still a student he designed a "hydraulic computer." It used a water wheel (made from a bicycle wheel) as the clock and various plastic connectors (designed to shunt water to a new path in regard to logic switches he designed) and it worked![63]

Johnson would later join the faculty of SIUC, teaching computer programming; he would become one of SIUC's first "webmasters." Along with his computer, design, and geodome know-how, Johnson was also an accomplished aviator.[64]

Although not a founding member, an early recruit was Judy Ashby. A Carbondale resident since 1970 and the possessor of an MA from SIUC, Ashby was a Bucky fan. She saw him lecture on campus years earlier when he spoke at the Newman Center. Ashby says, "I squeezed into the back and I think I was there for four and a half hours. It was standing-room-only."[65]

"Always aware" of the home, Ashby became involved one day after she happened upon Bill Perk drawing the World Map on some cement at Turley Park as part of a

The loft, *c.* 2003.

presentation. She struck up a conversation. Perk had just purchased the Dome.[66] Ashby volunteered herself and found that her first job was to clean the Dome's interior. She recalls, "It took a couple of days to fully clean. It was VERY dusty." Yet, despite the dirt, Ashby found it invigorating. She says, "It was great to actually be scrubbing the floors where the man actually walked!"[67]

Carbondale businessman Larry Weatherford was another early recruit. A native of Champaign, he came to SIUC in 1966. After starting out in liberal arts, he segued to the art department since that was where the "most interesting people" were. One of those "interesting people" was, of course, Buckminster Fuller. Weatherford once sat in on a Bucky lecture at Brown Auditorium.[68]

The Weatherfords lived away from Carbondale for a time, but they returned in the 1990s. Before recently selling his company, Weatherford ran SignGang. Active in the community, Weatherford knew Bill Perk and others involved with the Dome and they were happy to welcome him into the fold.[69] Weatherford remembers early meetings of the board taking place inside the Dome until his concerns about black mold within the structure demanded that the meetings be moved.[70] Despite that issue, other aspects of the home were still operational. He says, "I was pleased that the heating system was working! Concrete and some metals don't care for each other after a while."[71]

One aspect of the home that Weatherford has always been intrigued by was the fence. However, as the home endured, the fence continued to fall apart. As pieces began to break and fall to the ground, the slats were collected and put into storage for eventual reassembly.[72]

Like so many other board members, Larry Busch was an SIUC grad; he has a bachelor's degree (1969) and a master's degree (1970) in design. In his college years, Busch took classes taught by Bill Perk and attended lectures by Buckminster Fuller. Once, he even got to have lunch with Fuller himself. "I was always impressed that he remembered my name. He called me 'Larry,'" Busch says. Later, Busch taught in the Design Department and served as the department's chair. After being recruited by Perk for the Dome group, Busch remembers his early days on the board and the constant spit-balling of fundraising ideas. "We were constantly talking. Sometimes preposterous things—like maybe Ben & Jerry will make Bucky-flavored ice cream!"[73]

Other ideas on behalf of the Dome have sometimes been equally fanciful. At one time, early in the Dome's new ownership, the idea was floated about picking up and moving the home off of its Forest Avenue lot and to another location, perhaps one closer to the city's center. Fortunately, this idea was discarded as it would have ruined future opportunities for preservation grants.[74]

The most important step for the group, however, occurred on December 24, 2002. That date marked the official donation of the Dome from Bill Perk to the NFP.[75]

Another major step occurred on October 1, 2003, in a meeting of the Carbondale City Council. That night, the Council took up recognizing the Dome as a local historical landmark. Held in SIUC Student Center, the Council's discussion was one of two items of "new business" to be discussed that evening. Yet discussion was short. Following the presentation of a report on the topic by City Senior Planner Lisa Reime, the Council voted unanimously to recognize the Buckminster Fuller Dome Home as a local landmark.[76]

Reime, now Lisa Koerkenmeier, was an SIUE grad who had joined the city of Carbondale as a city planner in 1997. She admits today that she was not overly familiar with Buckminster Fuller when she first came to town. However, the enthusiasm of a "very persistent" Bill Perk (who was often seen at City Hall) proved contagious, and soon, she too was a *de facto* friend of the Dome.[77]

About a month after the council meeting, a ceremony was held at the Dome, where a plaque noting the home as a bone fide local landmark was installed. Despite overcast skies that day, the mood on site was upbeat. Joni Reeves said of the honor, "This is a stepping-stone for us to apply for national historic landmark status. Both statuses will allow us to apply for grants that we will need to renovate the home."[78]

Meanwhile, the first application to the city of Carbondale to turn the residence into a museum was filed by the NFP on November 11, 2003.[79] Cornelius Crane said at the time:

> I wanted to make sure we had all the legal foundation—such as permission from the city council to create a home museum—and other obstacles out of the way before we started grant writing…. I hope we get some grant award money and get some wind in our sails for this.[80]

On January 20, 2004, the City Council had it on their agenda to vote on the museum resolution. It passed.[81] The RBF Dome NFP is the first home in Carbondale to receive museum status.

The Dome's front yard, with the dormant fountain in the foreground and wobbly fence behind. Note the stack of fence planks lying on the grass in the background. (*Courtesy: White & Borgognoni*)

From left: Board members Bill Perk and John Johnson; local resident Bill Ittner; Carbondale city official Lisa Reime; local resident Dede Ittner; and former NFP President Cornelius Crane.

A few months later, another milestone—of sorts—arrived. In mid-2004, the Dome was named the second "most endangered" historical structure in the state of Illinois by the Landmarks Preservation Council of Illinois (LPCI). Though nominated by the NFP in order to obtain awareness for their efforts, the recognition was still a bit dubious. NFP President Crane said then, "It's a select group, so we weren't sure we'd get on the list. There are two or three properties [who also applied] that apparently have wrecking balls sitting in front of them."[82]

Cornelius Crane told the *Southern Illinoisan* at the time, "This is the first major recognition we've gotten. It's kind of a weird honor—being recognized because the house is deteriorating."[83]

The benefits of being "endangered" though happened almost immediately. To push the recognition along, board members traveled to Springfield for a banquet held by the LPCI. There, they met the chairman of the group. Perk recalls:

> Lo and behold, he said, "You guys have to have an architectural evaluation done to get going on fundraising." He promised us $5,000 and we said "Can we get that in writing?" We were happy to have had that publicity but getting $5K was totally unexpected and wonderful! Our next step was to hire someone appropriate.[84]

Perk and company acted quickly on that recommendation. In 2005, they retained local architectural firm of White & Borgognoni (W&B) to do an assessment of the Bucky property.

The finished report was necessary, thorough, expensive, and brutal. It also took some time to complete. Begun in March 2005, it was not submitted in a final form until February 2006. The finished report, complete with photos and architectural drawings, was also long—over 150 pages.[85]

As any would-be homeowner knows, a home inspection prior to buying is *de rigueur*. Often, these reports tell you more than you want to know. Though the board had long been stocked with architects and builders, only so much was plainly visible. A detailed review brought new issues to the fore. Among the report's "highlights" are the following:

> the sliding glass doors and frames had the following issues: possible water infiltration, missing sealant, missing screens, the need for the removal of non-original hardware; wood deterioration, and were not watertight;
>
> the four windows of the home had the following issues: they were non-operable, had broken panes of glass, and were not watertight;
>
> the driveway and sidewalk around the house showed innumerable cracks;
>
> the doors of the home had the following issues: dirty scuff marks, blemishes, marks from removal of non-original hardware, and lower portion deterioration;
>
> the concrete floor slab was exposed in many places; some floors have been painted (the living room floor, for example, was noted as being green [?], the bedroom floor was red);
>
> finally, they noted, if the home was to become a museum, it would also have to be made accessible to the public per the Illinois Accessibility Code and the Americans with Disabilities Act.[86]

Photo of the Dome's (not original) front door, *c.* 2003. (*Courtesy: White & Borgognoni*)

Additionally, review of the home's eleven overhead structural polygons found them to often be "water damaged" and in a condition from "fair" to "poor."[87]

The report also made note of the myriad of places where latter-day fixes took the Bucky Dome away from its originally built state. For example, the refrigerator in the home had long-ago been replaced. It was determined, however, that the stove in the home was probably the original article.[88]

Some assessments, though, got tricky. Since it was considered completely uncouth to take photographs of bathrooms in the 1950s and '60s, there is little evidence of how those rooms originally looked. When the investigators got to the bathroom floors, they had to bring in Ira Parrish to try to tell them what was originally laid down.[89] Both Parrish and one-time owner Mickey Mitchell believe that cork was the original floor covering.[90, 91]

Even today, as the restoration moves towards its finalization, some refurbishing takes a little guesswork. Though restorers know that the medicine cabinet, tank, and bowl of the bathroom and its approximate 4 × 4 ceramic tiles are original, what about the "vanity" lights and towel bars and wall mirrors? For now, until proved otherwise, the tiles will remain in place. Two sets of "vanity" lights wired and hanging above each bathroom sink are known to be later additions, however, and they will be removed.

Photo showing one of the dilapidated sliding doors of the home.

Looking into the bedroom. Note the presence of old blue paint.

As the Dome's later contractor Blair Wolfram has pointed out, the job of those working on the project has been to get into the thinking of Buckminster Fuller and "how he saw the future of housing in 1960." At the same time, the restorers have had to be loyal to the original design of the house—even to its flaws. For example, the guttering system Bucky originally designed has, over the years, proved less than ideal. According to Wolfram, "Mostly those narrow water-removal railings that circle the roof just gather leaves and dirt. Bucky would probably have redesigned that at a later time."[92]

Something else Fuller would no doubt have rethought was the home's single external electrical outlet. Strangely, the outside outlet was positioned directly under one of the gutter spouts, right where rainwater would run right down over it. Today, for safety, this outlet has been deactivated.[93]

Full restoration has also meant the loss of some comforts. The wall unit air conditioner in the bedroom that got renters through many a hot summer would have to be—and has been—excised.[94]

Then some things are just gone. Of the wood that makes up the vertical trapezoidal sides of the home, only one panel—the north/bedroom trapezoid—was fully intact. The other trapezoids around the building are made up of some original panels and some replacement ones.[95]

Not everything from the W&B report, however, was bad news. The underlying foundation of the home was pronounced to be in good shape, and some places just needed a good scrubbing.[96] It should be stated that not everything reported was unexpected; any home in its fifth decade, and having had multiple tenants, is going to show its age.

The report also offered various recommendations as well as various options for the home's rehabilitation. Of course, one of these suggestions was the extreme suggestion that the entire home be dismantled and carted away (which, they noted, would also require the removal of the remaining fence) and then be inspected, piece by piece, before being reconstituted either on the site or on a new site for later transport back to the original lot.[97]

Though recognizing the need for the report, Thad Heckman questioned a suggested dismantling of the Dome especially since the report did not address a primary issue particular to domes—its current adherence to "sphericity."[98]

A dome gains its strength by its internal equality of forces and radial symmetry, and the report did not address this central question: just how far out of geometry was the Dome? Heckman explains, "Conventionally framed structures are relatively easy to verify if they are 'square, level and plumb' by simple vertical, diagonal and right angle measurements. Not true with a geodesic dome." Therefore, to determine this relationship, Heckman employed a 3D laser level—a Hilti PM 24 multi-axis laser plumb device and a Hilti laser range finder—and his own university intern, Kelly Speckhart, who, with the help of the software program Sketch-Up, created a 3D computer model of a dome in digital space. The purpose was simple: to create an "ideal" geodesic dome against which to compare the current measurements of the actual *in situ* Bucky Dome.[99]

Working at the Dome, Heckman placed various "witness marks" on the floor. Using laser tools, he was able to take various measurements of the Dome's vertexes from the floor, including measurements to the Dome's highest vertexes, places that could not be reached easily even with tall ladders. Once this was done, the results

The broken bedroom window, "fixed" with duct tape.

could then be compared to the computer model. If the actual Dome's measurements were all equal or even close to the computer model, then the Dome was in its proper shape. If they were not, then work had to be done to bring the Dome back to its true spherical nature.[100]

Heckman's findings discovered that some parts of the Dome were not symmetrical, including areas that were not visible. Yet while some of the vertexes were obviously out of alignment due to ill-conceived repairs, general wood rot, and the like, a detailed investigation revealed something else that was not obvious—that the Dome was actually still mostly in its appropriate shape.[101]

The findings showed that various areas—the north side—of the Dome for example were quite close to the computer model. This yielded insight to the notion that the interior walls and framing for the loft also served to stabilize the overall Dome itself. Knowing that the northern portion of the Dome was not only stable, but that one vertex was exactly the same as the computer model and many others were very close meant that the Dome did not have to be disassembled and rebuilt at all.[102] Armed with this information, Heckman contacted Wolfram, who also agreed that dismantling the Dome was unnecessary. This decision had a significant impact on the preservation and, of course, on the budget.

Right: Photo of home's upper, west-side loft area.

Below: Photo of home's interior and east-side loft area.

Later, it was learned that the cause of the home's "dimpled" appearance was an incorrect earlier repair. About twenty years after the home was built, eighteen out of sixty of the home's original panels had been replaced and others had been modified. While a valiant effort was made by whoever undertook the fix, a small mathematical error ended up causing epic trouble.[103] As contractor Wolfram explained later, "The replacement tiles [triangles] were cut to the correct length but the wrong pattern. They didn't adhere to the Pease originals and that caused that part of the roof to slope inward." This has since been corrected.[104]

Though the NFP and the Dome itself seemed to keep turning up problems, in the early 2000s, the group was bolstered by a new explosion of "Bucky Love." In 2004, the U.S. Postal Service issued a commemorative Buckminster Fuller postage stamp. The 37-cent stamp co-opted the 1964 *Time* magazine illustration of Bucky with his head transformed into a geodesic dome. To celebrate the stamp's release on July 13, 2004, the NFP (with the permission of the U.S. Postal Service) turned the Bucky Dome into a one-day-only U.S. Post Office.[105]

Then NFP President Cornelius Crane said, "We did all sorts of things to try to raise money. Little fundraisers. Innovative little things."[106] One of the later, post-post office events was a combination art gallery opening and silent auction. Local artists donated works and local businesses donated goods that could be raffled off. One of the items up for grabs was a special Buckminster Fuller print donated by the Carl Solway Gallery of Cincinnati.[107] Joni Reeves adds, "We did everything short of bake sales! And we were always trying to get ourselves into the newspapers."[108]

The Bucky Dome got some major recognition in February 2006 when it was added to the National Register of Historic Places by the U.S. government.[109] Not to be confused with federally recognized National Landmark status, which is different, being added to the National Register is usually an important stepping stone to that other major honor. Established in 1966 by the Historic Preservation Act, the Register is administered by the National Park Service. There are, currently, over 1 million locations and buildings on the National Register.[110]

The initial application for the National Register came via the White & Borgognoni assessment; they attached an application as an appendix to their 2006 dossier.[111] At the time of the application's authoring, it was noted that the Bucky Dome was eligible for the Register under several of the provisions: Criterion "B," for national significance based on its association with Fuller; Criterion "C," for local significance as an example of a geodesic dome house; and Criterion "G," for properties with exceptional importance that have achieved significance within the last fifty years.[112] As with the earlier local designation, being named to the Register allowed the Dome to now qualify for various other grants and assisted them with various tax-related concerns.

As the 2000s percolated along, it brought along more interest in the Dome and its original owner. In 2007, the Illinois Historic Preservation Agency came out with its list of 150 architectural wonders in Illinois. Southern Illinois was well represented and, to no one's surprise, the Dome made the list along with the Stinson Memorial Library located in Anna and the Cairo Custom House.[113]

In 2008, the Whitney Museum in New York City staged a major Buckminster Fuller retrospective. The exhibit, *Buckminster Fuller: Starting with the Universe*, was open to the public from June 26 to September 21, 2008. This expo culled original documents, notes, illustrations, and models to give a full understanding of Bucky. Along with ample press coverage, the exhibition was also featured on CBS' *Sunday Morning*. After running in New York, *Universe* traveled to Chicago in 2009.[114]

Part of the Chicago exhibit included a film presentation produced by SIUC's own in-house film company, Barking Dawg Productions. The video, a production of marketing department chair Terry Clark and various students, blended both archival footage with interviews with people influenced by Fuller. Kyle Tezak, a 2008 grad who worked on the film, said about it, "It's not necessarily an informational piece; it's more of an introduction to the exhibit. [It aims to capture] Fuller's presence and to explore his influence."[115]

Producer Clark added, "For a lot of us alums of that era, Bucky Fuller was an icon. It was like having Elvis on campus."[116]

In 2006, a CD was put out. Titled *Roam Home to a Dome*, the twenty-six-track recording consisted of musical tributes to Bucky by various artists; spoken word passages featuring Bucky himself; and audio recollections by those who knew him. Most of the audio was recorded inside the Dome; the building has always been revered for its acoustics. Under the guidance of Cornelius Crane, the CD was recorded, mixed and mastered by Michael Lescelius; its art direction was by Stephen Gariepy.[117]

The finished $15 CD was sold locally with all proceeds going toward Dome preservation. A limited-edition version, for $40, came with a fold-out tetrahedron. Some of the cuts on the album include "Everything is Round," "What If We Are One," and "Wreck of the Dymaxion Car."[118] The album's liner notes stated:

> In the Spring of 2005, dozens of local and nationally recognized musicians began gathering in R. Buckminster Fuller's acoustically unique Dome home to record music inspired by Fuller's life. The fruits of their collaboration, *Roam Home to a Dome: A Benefit Compilation*, has been released.
>
> Unlike many "intellectuals," R. Buckminster Fuller lived in a world of people, places, things and action rather than the antiseptic sanctuary of the "mind." Mr. Fuller was a thinker no doubt. But as he liked to say, he did it "out loud." As artists we'd like to show our appreciation to him for changing how we think. With this album we do it OUT LOUD. —Mike Lescelius.[119]

For Kathy Livingston and her husband, Paul Matalonis, the CD was, largely, their introduction to the NFP board—a board they would soon join. Livingston grew up in Springfield, Illinois, but attended SIUC, graduating in 1996. During her student days in Carbondale, she lived on Forest Avenue and frequently walked by the Bucky Dome. Though not overly familiar with Fuller during her undergraduate days, Livingston learned about him later from various Fuller-infatuated friends, like one-time Bucky Dome renter Deb Browne.[120]

Along with both being social workers based in Carbondale, Livingston, and Matalonis are also active in the local music scene. It was via their friendship with Cornelius Crane as well as various other *Roam* talents that they first got involved in the CD project and then with the NFP.[121]

The day that Livingston and her female vocal trio, For Healing Purposes Only, recorded their tracks for the CD was her first time inside the Dome. While her group

Cover artwork of *Roam Home to a Dome* benefit CD.

sang in the home's living room, engineer Mike Lescelius was headquartered in the bedroom. One of the songs Livingston's group did that day was John Denver's song "What One Man Can Do," a song Denver wrote about Buckminster Fuller.[122]

Livingston found the day's experience transformative. Along with the Dome's great acoustics, she says she was struck by "the quality of the light [in a home] without any rectangular walls. It was inspirational. I was so enamored. So many people were there that day talking about Bucky, and to be able to learn about him while being in *his* Dome!"[123]

Livingston would be on the board beginning in 2005. Since then, she has become so "Bucky-fied" that she even named her current performing duo after one of Fuller's famous talking points; her musical partnership with Curt Wilson is named Meridian 90.[124]

To publicize the *Roam* CD, in December 2006, the NFP sponsored a float in Carbondale's annual Christmas parade.[125] For their float, the NFP assembled a mini, metal geodesic dome, adorned it with twinkling white lights and put it on a flatbed trailer further covered in decorations. From an onboard boom box, they blasted the just-made CD. Since this mini-dome's first appearance in that parade, it has been reemployed at various events. In later years, it has sat on the Dome's lot as a decorative, protective covering over the yard's inactive fountain.[126]

Other board fundraisers over the years have included the following:

In 2007, the book *Carbondale After Dark* by former Bucky Dome renter H. B. Koplowitz was reissued, a portion of its local sales were donated to the NFP.[127]

Beginning in 2008, Orlandi Vineyard in Makanda, Illinois, started hosting an annual Bucky benefit. The vineyard's owner, Gary Orlandi, had first heard of the not-for-profit

Shot of the oft-used mini-dome sitting in front of the Dome. Here the smaller dome is draped with Christmas lights.

on the radio and decided to help out. The Sunday afternoon "Vine-Maxion" event included not only wine tasting but participation from various area businesses including Holistic Healing Arts, Advance Energy Solutions and the Town Square Market. The Ivas John Band provided entertainment that first year with proceeds going toward the NFP.[128]

The following year, the Orlandi occasion got even bigger. Held in June of 2009, it included silent and live auctions, a raffle, live music, food from Mase's Place of Pomona, and a return of the Ives music outfit with special guest Candy Davis. The day also included, one assumes, ample amounts of vino. About 500 people attended the second annual event and the day yielded about $5,000 for the NFP's coffers.[129]

Then, just in time for Christmas, in 2009, the organization offered for sale mini-geodesic dome models perfect for your desk, coffee table or rumpus room.[130]

Still, despite warm and fuzzy feelings and wonderful steps forward, after almost a decade of hearing about the Bucky Dome's restoration, some citizens were growing weary with the pace of the progress. As one board member once stated, "When you are a board member for the Bucky Dome, you learn that people want it done yesterday."[131]

In May 2008, the *Southern Illinoisan* published an article about a resident of the Arbor District who expressed aggravation with "the ardently slow process" of the Dome's repair. In an e-mail to the District's neighborhood association, the resident wrote, "I was wondering if there is any projected timeline on the repair of the dome," noting that the plastic-covered protective dome in their neighborhood was becoming a "significant eyesore." "At what point," the author asked, "will it be determined that the dome can be restored, or will fundraising efforts continue with no end date in sight?"[132]

Southern Illinoisan sought out Cornelius Crane for a response. "Our only restriction on time is money," Crane said, noting that a major portion of the monies so far collected to save the Dome had to be dedicated to such things as property taxes and basic maintenance. Crane also noted that several other hurdles still stood in the group's way. For example, obtaining National Landmark status could only be achieved after a structure was at least fifty years old. At that time, the Dome was only forty-eight.[133]

Only a few days after the original article appeared, *Southern Illinoisan* published another letter to the editor. Sent by two other Arbor residents, they spoke of their support of the Dome restoration, no matter how long it took. They said, "We are moved by these nearly six-year-old efforts in building and in fundraising by our friends and neighbors, Arbor District and otherwise, who we see on the grounds ... laboring for something they believe in."[134]

After the original Arbor District communiqué was relayed to the NFP, the foundation sent letters to all neighborhood residents noting the soon-to-happen removal of an unsightly tree stump on the property and the upcoming replacement of the sheathing over the protective dome.[135]

Actually, the Arbor District has, by and large, always been very supportive of the Dome. When the neighborhood used to have its annual block party/festival, it always invited participation from the NFP who frequently staged a concurrent open house. One year, at the event, the NFP offered "Dymaxion sundaes," an ice cream treat consisting of orange sherbet and dark chocolate. The sundaes were served in an edible cone—to avoid waste.[136]

In 2009, the NFP filed their application for the Dome to become a National Historic Landmark.[137] Though the terms "national landmark" and/or "historical site" are often bandied about interchangeably, the honor is actually given out rather rarely. Since the program was established by the National Park Service in 1960, only 2,532 locations in the entire country have received the honor. So far, some of the structures to be given the designation include the St. Louis Arch and the Alamo in Texas.[138] One reason for the relatively few truly historic places is the stringent application processes that have to be gone through in order to obtain the official sanction. In the case of the Dome, with the restoration process underway, NPS guidelines regarding preservation, restoration, and reconstruction have to be followed to the letter. These include returning every aspect, every detail, of the structure back to its original state or as close as can be achieved.[139]

Furthermore, grants also demand, for locations like the Bucky Dome that are currently undergoing restoration, that all work plans and architectural documents have to first be submitted to the federal government before a single nail can be nailed or a single nail can be removed. These edicts—though necessary and beneficial in the long run—can nevertheless add months, or even years, to the length of a restoration project.[140]

Nevertheless, despite the long application, it has always been the plan of the NFP to have the Bucky Dome declared a National Historical Landmark. The honor brings with it not only prestige but an extensive set of legal protections that would help preserve the Dome possibly indefinitely.

It was the sight of the over-dome's lingering plastic—frayed and often seen waving in the breeze—that first spurred Janet Donoghue to get involved with the Bucky Dome. An SIUC student, with a BA in theater and an MA in speech communication, Donoghue first visited the Dome during some of its rental years, thinking at the time "this is a strange building."[141]

Donoghue seemed primed to become involved with the NFP. She was, at the time, not only working on her Ph.D. at the university, but also hosting a local radio show, *Greenhouse Rebellion*, an environmental program over WDBX, which she began in 2005. Often focusing on such Bucky-type topics as sustainability and the environment, Cornelius Crane had once been one of her on-air guests.[142]

After Donoghue read the original complaint letter about the Dome in the newspaper, she decided she wanted to be one of the people to finally get the plastic off of it. So, with a newcomer's zeal, and she says, worn out from her own "Ph.D. grind," Donoghue attended her first NFP meeting where she was quickly welcomed into "Bucky Land." Having worked for a time in the Chicago area as a development officer, Donoghue was the first to write a formal development plan for the group. Later, she'd do some early grant writing and create the first Bucky-oriented "camp" for kids.[143]

All of Donoghue's work eventually paid off—for both the organization and for her. In 2010, she was hired to act as the NFP's director of development. With the exception of some hired-help for the board's webpage, Donoghue remains the only paid employee the NFP has ever retained. Originally, Donoghue worked from home but, later, she had a more formal office when the foundation finagled a short-term workspace at a local artist collective.[144]

Donoghue would serve in her Dome position for, approximately, a year and a half. She says today, "I really made Bucky and the Dome the center of my life. I was one of *those* people."[145] Just before Donoghue's joining, the board also welcomed Brent Ritzel.

Ritzel was a Carbondale native. Though the son of an SIUC professor, Ritzel grew up not really aware of the Dome until he was in high school and went looking for a project for the school history fair. After taking on the subject of the Hundley House murder case the prior year, this time, Ritzel was directed to the Dome by his parents. Ritzel's resulting project—with an accompanying research paper written by classmate Kristen Fligel—was a scale model of the Dome that took over 150 hours to complete and included 5,000 independently cut squares of sandpaper to replicate the house's shingles.[146]

Ritzel researched his project by reading up on Bucky and then by going and knocking on the door of the Dome and asking then tenants, Pete DePaoli and Kim Alexander, if he could come in and look around.[147] Ritzel still remembers his first impressions of the Bucky Dome. He says, "I remember the difficult angles and the odd ways that the walls hit the ground. I mean it was a whole different geometry of living—it would have been like living in a different reality."[148]

Ritzel did well with his history fair project and it even went to the State championship. Sadly, even though the school retained the model, a few years ago, it was accidently knocked off the shelf, and much to Brent's dismay, the school discarded it.[149]

After high school, Ritzel attended Northwestern and graduated in 1990 with a degree in philosophy. Later, he obtained an MA in public administration. After living in Chicago and Denver, Ritzel returned to Carbondale where he went to work for a renewable energy company.[150]

Brent Ritzel poses alongside friend and Dome donor Tootie Wesslemann inside the Dome.

By 2008, Ritzel owned equity in a local energy company, Equitech International, where one of his fellow shareholders was Bill Perk. Perk invited Ritzel to the next meeting of the Dome board and Ritzel accepted.[151]

Though anxious to get involved, Ritzel was originally surprised and alarmed by the then state of the Dome. "It was a pit," he says. "It looked like a dilapidated construction zone even though there was no construction going on." Along with all the inside issues, the outdoor fence was now a complete eyesore and a hazard. Ritzel says, "We were risking major liability with that thing. There were exposed nails. Someone could have been impaled."[152]

Despite being daunted by what he saw, Ritzel threw himself into the cause. He was soon devoting thirty hours a week to the Bucky Board. He says, "I guess I was inspired by the fact that this was a group with just one very clear mission: restore the Dome. And, to do that, all we had to do was one thing: raise money."[153] Other board members took notice and fast-tracked Ritzel's membership and his ascent to be the group's president.

Though Cornelius Crane had admirably served both the group and the Dome since its earliest days, he was ready to move on and welcome in a new generation.

Close-up of the Bucky fence and one of its short-term patch jobs.

Ritzel's enthusiasm made him a perfect new leader. He was elected NFP president in February 2009.[154]

In his new role, Ritzel adopted for the group a higher, more aggressive profile. He says, "We wanted to make sure that we always had a stand at any community event. We wanted to create a constant public presence."[155] Along with local projects, Ritzel also pushed the board to take part in other events within the larger southern Illinois and eastern Missouri area, for example, taking part in St. Louis' annual Earth Day events.[156]

One of Ritzel's early goals was to educate the community not only about the cultural importance of the Dome but also the potential economic advantages of it as well. Ritzel commented: "A popular historic site like the Bucky Dome could bring in two to four million dollars for the local economy and help with job growth as well."[157] Ritzel also did what he could to change some attitudes among board members:

Before, I think, everyone was always depressed by the thought of matching funds for grants but we had to get beyond that. I told them, among other things, that we could monetize all of our work, our labor on behalf of the dome and that would always count against grants.[158]

With fellow new member Janet Donoghue, Ritzel put together the group's first plan for fundraising. He also began to scour Google for potential grants and other awards that the house might be suited for. He knew he had something when he came across the Save America's Treasures (SAT) program.[159]

Now discontinued, the SAT grant was a program administered by the National Park Service in conjunction with the National Trust for Historic Preservation, the National Endowment for the Arts, Heritage Preservation, and the National Park Foundation as well as the President's Committee on the Arts and Humanities. An SAT grant, if awarded, was worth $125,000.[160]

Ritzel and Donoghue did the primary writing of the SAT grant proposal and, as Ritzel recalls, submitted it "about thirty seconds before the deadline."[161] While the SAT application moved forward, Ritzel kept applying for other funds. He also helped soothe any residual neighborhood concerns, mainly, he says, by taking down completely many of the more unsightly parts of the Bucky Fence.[162]

After the removal of most of the fence ("an all-day job," according to one volunteer), the Dome sat exposed and visible from the street for the first time.[163] Security for the structure would have been a concern had it not been for over-dome.

Two other new faces came to the board in 2009. One was the late Linda Hostalek, a local physician, who would remain affiliated until 2011 when she moved to Hawaii.[164] The other was architect Thad Heckman.

Born in Mexico, Missouri, and raised in Paris, Missouri, and west central Illinois, Heckman later pursued a design degree at SIUC. He graduated in 1979, one of the last thirteen students turned out by the "old design school" of Herb Cohen, Bill Perk, and Bucky Fuller, before it was absorbed into the Department of Art and Design. Having first learned of Fuller when he was still in high school in Payson, Illinois, Heckman got to meet the great man himself when he was in his senior year in Carbondale. Heckman recalls, "I think, when I met him, I just stood there with my jaw hanging open. I'm mad at myself now. I could have asked him anything—but I was too timid."[165]

Heckman had known of the Bucky Dome since college and got to go inside the Dome for the first time in the late '90s. Heckman's first impression though was a bit underwhelming:

> I remember it being kind of junky inside … I had attended my first year courses in the old Design Department geodesic domes—I loved being in them and we would remark that time seemed to pass differently inside them. I was familiar with domes and I was surprised at the condition of Bucky's dome.[166]

Heckman, by this time a professor at SIUC, became more formally involved with the Dome's restoration when, in 2003, he was approached by Cornelius Crane to help with the original Carbondale Historic Preservation District nomination report. Heckman was asked to possibly write a study on the Dome's full restoration. Heckman did some work and quoted the board around $7,000 to finish the write-up. Yet, for reasons he still does not know, Heckman was passed over and White & Borgognoni were hired instead and at a substantially higher fee.[167]

At about that same time, Bill Perk offered Heckman the chance to work out of the Dome. Believing it would be good to have some sort of occupancy in the Dome and

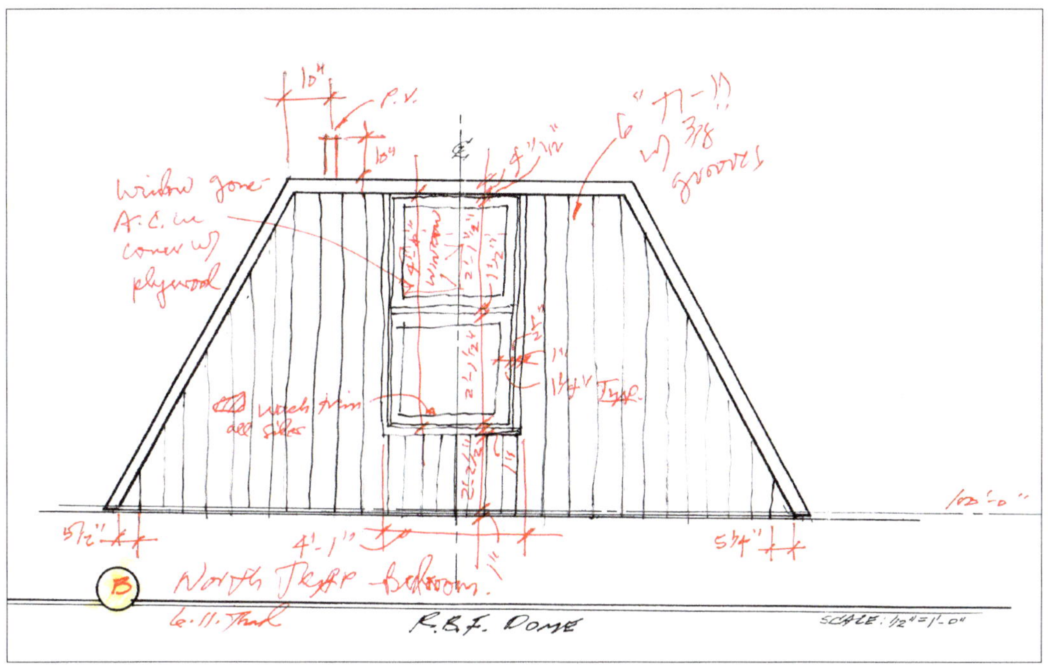

Thad Heckman's early field notes and measurements for the Dome's north-facing side.

recognizing the synergy of the having an architectural firm headquartered there, Perk made the suggestion. Ultimately, however, too many various issues prevented any sort of move.[168]

Still, despite that, some years later, around 2011, Heckman was approached by Brent Ritzel about joining the NFP. Heckman joined and was a member until late 2017. Heckman would soon assume the (much needed) role of lead architect on the preservation.[169]

One of Heckman's first undertakings was to organize the actual restoration of the home. He mapped out a plan and broke it down into three phases:

Phase I: Stabilize and repair the main structure of the home; preservation/restoration of its appearance to match its original design;

Phase II: Repair of the inside of the Dome; preserve/restore it back to its appearance during Bucky's residency;

Phase III: Repair/preserve/restore the home's outdoor landscaping, fence and in-ground water feature.

A Phase IV has since been added; it calls for the home's original interior furnishings and other décor to be recovered or matched as accurately as possible.[170] The three phases were seen as a way to better communicate fundraising efforts to the public and to better order the work. Heckman adds, "Most importantly, I put the exterior preservation and protection as the highest priority—hence Phase I."[171]

Along with establishing renovation "phases," Heckman also began work on the detailed architectural drawings needed for Phase I; with or without money, work could not proceed without a detailed map as to how the finished Dome was to look. Heckman's records indicates he began this work on February 21, 2011.[172]

Another of Heckman's early bits of business was arranging for an asbestos evaluation. That review was done by KAM Solutions of Mattoon, Illinois. KAM is Kent A. Metzger. Metzger was kind enough to donate his services.[173] To everyone's relief, Metzger's found that the Dome was surprisingly asbestos free. Originally, everyone feared the worst. At the time of the Dome's original construction, building elements were often loaded with asbestos containing materials (ACM's), usually in order to be fire retardant or as a binding agent. However, the Dome somehow got spared and, at the time of its evaluation, was found to have only minimal amounts of ACM and these amounts were found to be "non-friable," meaning they did not release airborne fibers that could, potentially, be inhaled.[174]

Still, the small amount of asbestos that was found did require removal and disposal. Fortunately, the amounts were so small that restorationists were spared the need for "space suits" and other equipment often needed in the exhumation of asbestos from vintage properties.[175]

In 2008, as noticed by Donoghue and others, the outer plastic shell of the over-dome was becoming frayed, the victim of its battle with the elements. Thankfully, though, the underlying steel of the protective outer dome was still in good shape.

At first, the NFP approached various companies about crafting a replacement tarp. A tent company in Indiana wanted over $28,000 for a new covering. Finally, the group hired the outdoor shelter company Farmtek, headquartered in Dyersville, Iowa. From them, the NFP purchased endless yardage of a woven greenhouse material that was industrial in strength and UV deterrent but still translucent. Farmtek then sewed huge panels of the material together, creating a quilt large enough to cover the Bucky Dome. Farmtek shipped the fabric to Carbondale where it was placed over the outer dome.[176]

Unfortunately, once installed, it took little time for the seams of the fabric to start to separate. Soon, panels were dropping off and gathering in coils around the base of the Dome.[177] Finally, the NFP appealed to Cincinnati's World fx, Inc., the leading company in custom-manufactured globes.[178]

In business since 1995, World fx is the creation of brothers Bill and Todd Ulrich. They provide oversized globes of nearly every kind and size to museums and festivals around the world. The brothers first encountered Buckminster Fuller not long after founding their company when they created globes for a Dymaxion projection project organized by the Fuller Institute in California. The event coincided with Bucky's centennial. While in California, the brothers made the acquaintance of Bucky and Anne's daughter, Allegra Snyder, and of Bill Perk.[179]

Perk informed the brothers of Carbondale's Bucky Dome and its need for a new covering. Remembers Todd, "We stopped in Carbondale on our way back to Ohio and Bill gave us a tour of the Dome. We picked up the tarp and took it back with us."[180] Rather than take the old fabric back to their Cincinnati HQ, the brothers, instead, took it to their parents' farm north of Dayton. Todd says:

> I constructed on the computer a globe gore-generating algorithm that would allow us to cut the material into the gores like we would to construct a globe. A gore is a

The over-dome's metal skeleton stripped of its first plastic covering.

canoe-shaped panel, narrow at the top and bottom and widest at the "equator." With the algorithm I could cut the panels—around 36 total—into a size just slightly larger than the steel dome.

Then we marked the fabric and cut it out in that shape. We cut them just long enough to tuck under the Dome. The last thing we wanted to do [on site] was have to open it up or add more fabric. We sewed in a webbing and special strapping.[181]

According to Bill Ulrich, the sewing session took around fifty hours.[182]

The brothers returned to Carbondale on July 6, 2008, bringing the finished covering with them. Using a series of ropes and pulleys—an "ingenious method that Bill devised," according to Todd—the tarp was hoisted to the top of the over-dome.[183] On site that day were Bill and Todd, Bill Perk, and Cornelius Crane. Some trimming of trees and bending of chicken wire had to be done before the tarp was unfurled. Yet once it was on, "It worked great," Todd said.[184]

Also on the lot that day was videographer Christopher McCastle, who recorded the day's activities. Later, he edited together the day's highlights and added as background music an original composition by Stace England titled "Can't Stand in the Corner"; a sample lyric is "Inside the Dome, you can't stand in the corner no more."[185] McCastle's finished video is on YouTube.[186]

Street view of the Dome with over-dome in place after the majority of the Bucky fence was removed. Note that the covering of the over-dome seen here is its second, World fx-created sheath.

The final wrap job—practically worthy of Christo—covered every inch of the over-dome. At the end of the day, the men used a knife to cut openings for the Dome's door and windows. The Ulrichs also cut a two-foot wide cupola at the apex of the Dome for airflow.[187] The final cost for the new tarp was around $2,500.[188]

The replacement tarp could not have happened at a more important time. Only a few months after, Southern Illinois was hit by one of the most powerful storms in its history. Named El Derecho, the inland hurricane was a straight-line storm that hit the Carbondale area on May 8, 2009. Along with pummeling rainfall, winds were clocked at over of 100 mph. The deluge turned area roads into rivers and the storm's winds uprooted trees, downed power lines, and even overturned cars.[189]

As the storm raged, NFP board members had a long night of pacing in their basements, worried about their own homes and about the Bucky Dome. Though domes are designed to be among the strongest of structures, nothing like this had ever been experienced here before.

Though the Dome itself was protected thanks to the over-dome, and the over-dome was secured to the ground, this weather was unprecedented. If the over-dome got lifted up, it could be thrown around like a paper plate. Then, needless to say, if the over-dome blew off, that would leave the Bucky Dome fully exposed.

The morning after the storm, Carbondale and surrounding communities looked like a war zone. Over 3,000 trees had been toppled and limbs and other debris entirely blocked major roadways. As one utility pole fell, it pulled others down with it, creating a devastating domino effect. In some neighborhoods, entire sides had been pulled off of buildings, exposing them like giant doll houses. Even the city police station sustained damage.[190]

In the days after the storm, over 60,000 people in the area were without power. Even cell phone communication was affected as many cell phone towers were also felled. City officials reported thirty-four area buildings to be complete losses. SIUC later reported damages of over five million dollars.[191]

The morning after the storm, NFP members came to inspect the Dome. Many were fearing the worst. Remarkably, the recently attached covering was still intact. Todd Ulrich said, "What we did held. The corner of the dome was chained to a tree but it held. The storm knocked down a couple of trees in the house's yard though."[192] One of those trees was a 700 pounder that fell onto the protective dome. It made a part of the over-dome concave but did not damage the Dome underneath. When the tree fell, it dislodged the over-dome and slid it about 5 feet to the south. The over-dome then came to "rest" when it hit up against the house.[193] Nevertheless, despite the shift, the board considered themselves lucky.

Not so lucky, however, was Bucky's outlaying fence. Not surprisingly, the redwood barrier sustained the most damage with many slats having been blown loose and thrown wildly into the yard. In the weeks after the storm, volunteers gathered up the wood pieces and hauled them for storage to a (donated) warehouse belonging to Southern Recycling. Those pieces are still waiting there until the NFP is ready to rebuild the fence.[194]

In April 2009, despite all the actions taken toward the Dome's resurrection, the *Daily Egyptian*, published a story stating the Dome had just been condemned by the city of Carbondale.[195] Thankfully, the paper printed a correction/retraction a few days later as part of a longer story about the most recent activities of the NFP.[196] According to that April 22, 2009, article, over fifty former friends and colleagues of Bucky's, as well as current students and interested locals, gathered at the Italian Village for their regular brainstorming session. There, President Ritzel spoke of the home's future as a museum, one capable of attracting 20,000 people a year.[197]

As mentioned, with the Bucky Dome hidden behind plastic, to many, the home was the source of great mystery. Hence, in April 2010, to celebrate the home's fiftieth anniversary, the NFP launched the Fuller Dome Transformation Initiative and opened up the Dome to visitors. Though the tarp remained in place, visitors could wander inside the house and see what everyone had been talking about for so long.[198]

Along with tours, the fiftieth was also cause for a city-wide Bucky fest that included Carbondale proclaiming the week of April 19 as R. Buckminster and Anne Hewlett Fuller Dome Home Week.[199] Fuller-inspired events took place all over—on Main Street, at the town square (where a temporary geodome was erected), at Morris Library, and at the University's School of Architecture. Visiting dignitaries included Thomas Zung, Fuller geometry expert Joe Clinton, and World Game Institute co-founder Medard Gabel. At SIUC's Browne Auditorium, a one-man play, *Buckminster Fuller Live!*, was put on. A film by Noel Murphy, on the restoration of Fuller's Dymaxion Car No. 2, was screened at SIUC's Student Center.[200]

One of the event's special guests was Allegra Fuller Snyder, Bucky's only surviving child. She was asked to attend by Brent Ritzel; it was her first trip to Carbondale in forty years.

Appropriately, Bucky Week coincided with the fortieth anniversary of Earth Day on April 22.[201] "Earth Day" was another Fuller invention, begot in 1970. The festival was the precursor to the now annual "Dome Days" celebration.

The Bucky festival was also a financial success, raising over $17,000.[202] Yet even amid that success, the Fuller home was still scrapping for funds. It was an ironic situation as in many cases, more and more of Fuller's theories and concerns were, once more,

The launch day of the Dome's preservation was documented with this group shot.

being brought to the fore. Though the Illinois Department of Commerce and Economic Opportunity ponied up money in July 2010, the NFP was still on the lookout for both money and physical help for the building—even from its board members.[203]

Brian Gorecki (SIUC grad, class of 1983) saw Buckminster Fuller speak at SIU around 1981 or 1982 and was always "vaguely" aware of the Dome. However, he did not become involved with it until one day when he bumped into an acquaintance at a Dome fundraiser. When he realized he actually knew several people involved with the Dome, he, too, decided to get involved.[204]

Gorecki, might have gotten more than he bargained for. "At the time, the board did most of the [upkeep] work. We had work sessions. Anyone who had tools brought them over." Gorecki got to work on fixing one of the two imbedded hose-bibs on the side of the house; it was leaking. Yet, he said, "For a while it was more fence maintenance than anything else."[205]

Serendipitously, replacement wood for the Bucky fence was acquired. Gorecki recalls:

> The Bond Brothers [of Bond Brothers Construction in Ridgway, Illinois] had purchased an old redwood water tower—it was over 100 years old—from someplace in Southern Illinois and we were able to acquire some of its remnants.[206]

Board member Ed Cook remembers one weekend spent at the Dome where volunteers carried endless piles of "stuff" (assorted left-behinds from former tenants) out of the Dome and loaded it onto a donated truck for carrying to a donated warehouse: "It was nothing pertinent to the house, from what we could tell. But we saved it all, in case it did prove to have value."[207]

Cook, who had been a Bucky fan since he made a beeline for Carbondale in the early 1970s, got involved with the Dome restoration after bumping into Bill Perk around 1999 at a local restaurant. "I bought Bucky's Dome!" Bill announced to him.[208]

Over his career, Cook says he has constructed dozens of domes, some of which he has lived in himself. With his big Bucky interest, Cook seemed a natural for the board. He served on the NFP from 2010 to 2011. Cook also hoped of doing some of the restoration himself but concerns over a conflict of interest precluded him.[209]

The early 2010s also saw Shannon McDonald, SIUC assistant professor of architectural studies and interior design, join up with the board. McDonald has a master's degree from Yale. In the years just before coming to Southern Illinois, she became versed in all things Bucky after attending the *Starting with the Universe* exhibition in New York.[210] Once in Carbondale, McDonald became surprised that Carbondale wasn't doing more to highlight Fuller. Not long after that, she took her first drive by the Dome and, soon after that, she attended a meeting of the NFP.[211] Newly affiliated with the group, McDonald soon had her first tour of the Dome. She was surprised:

> I saw it before the start of the BIG renovation, when the covering and the plastic was still on … I was struck how it sits in this small-town neighborhood and yet it feels … commonplace. It doesn't feel awkward to be there or uncomfortable.[212]

Though she came to the home's interior, she says, with "no preconceived notions," she was nevertheless, "really intrigued by the inside. I was impressed with how everything was different but in scale. I think the home would be really interesting to live in, to experience."[213]

Also joining the board in 2011 was Benjamin Lowder. Lowder began at SIUC but graduated from the University of Illinois with a degree in communications in 1998. Lowder's introduction to the Dome came via his father, who, when he was a Saluki, lived on Forest. Later, on a nostalgic drive through Carbondale, Dad pointed out the famous house to young Ben.[214]

Lowder says he was always "academically aware" of Buckminster Fuller. His interest only accelerated when he decided to build his own home near Gerard, Illinois. For it, he decided that instead of home "status" he would craft a home in harmony with its surroundings—a thought in sync with Fuller.[215] Lowder's first board undertaking was updating the artwork for the NFP's marketing: "I re-drew the Dome. Looked at it from the top down. Deconstructed it. The first thing the new artwork was used for was the Dome fundraising capital campaign."[216]

Also an artist, Lowder's work is almost always based upon geometric shapes. Besides sculpture, he has worked in home design and photography. He has been exhibited at Art Basel. His fulltime job is at SIUE's Center for Spirituality and Sustainability—the site of Illinois's other Bucky Dome.[217]

In January 2011, a quantum leap in fundraising was achieved when the Dome was awarded a $125,000 "Save America's Treasures" matching-funds grant. Along with Carbondale's Bucky Dome, two other projects recognized that year were the National Cathedral in Washington, DC, and the Jacqueline Bouvier Collection in Boston. The awarding of this grant is usually a step towards recognition of the building as an official national landmark.[218]

Janet Donoghue, who worked with Ritzel on the grant, remembers the SAT news as one of the most important developments in the Dome saga. "It was our 'trim tab' moment," she says, echoing a Bucky phrase, "It helped us turn the corner and granted us legitimacy."[219]

While news of the SAT grant was wonderful, the mechanics of it is that the recipient must raise an equal amount of money in matching funds. If this does not happen, the original money is forfeited. Thankfully, matching funds can be tabulated via monetary donations or the donation of labor, professional services, or materials. For example, Thad Heckman's work as architect (his field work, research, and design) could be invoiced against the matching-fund debt. This would prove to be a considerable sum. The grant demanded a near-obsessive amount of documentation about the house's restoration.[220]

In 2012, Phase I working drawings (completed by Heckman earlier in the year) and other support documents had to be submitted to SAT, the city of Carbondale, and the state of Illinois, as well as the National Park Service.[221]

For some work, Heckman was assisted by Josh Rucinski, who, for a time, was officially recognized as the NFP's "chief volunteer." For Rucinski, his childhood took him from Minnesota to Montana to Illinois. Later, he joined the Army and landed in Germany. Throughout his life, Rucinski always had a love for drawing and carpentry. He is also a science fiction fan.[222] Could an interest in Bucky be far behind?

Later, Rucinski was a grad student in architecture studying with Heckman. Thad invited him to do some Dome work.[223] Of course, not much recruiting was necessary. Rucinski had an interest and was also looking to break up his schoolwork. Also, earlier, he had worked on a 3D project that dealt with domes.[224]

Rucinski's first trip to the Dome was during its over-dome era. When inside, he saw some of the vandalism that the Dome had endured: "The day I was first there, there was broken glass and beer bottles that had been thrown through the windows, on the floor. Someone had screwed shut the sliding doors to try to keep people out."[225]

The year 2011 also saw the launching of the NFP's webpage and its own Facebook page, both necessities in the digital age. Originally, the NFP's first website, however, only did the minimum, showing a few photos and listing a mailing address.[226]

Then, not long after its launched, as Janet Donoghue remembers, the site got "hijacked by the Russians" or, more accurately, got infected by a virus that was sourced back to the former U.S.S.R. The virus caused all the language on the site to be reprinted in Russian. It eventually fell to Donoghue to fix it. However, translating everything back to English gave Donoghue the chance to refocus the language of the website—in fact, of all NFP materials.[227]

> We had a real lack of continuity of message on the website and in other materials. I wanted to focus on how this was the only dome Bucky had ever owned, the only dome he ever lived in. It's exciting now to see my words used still, long after I left.[228]

Other developments in 2011 included the NFP's first dedicated office. Local businessman Mark Robinson donated use of a building he owned in Carbondale at the southeast corner of Oakland and Sycamore. Formerly a Coca-Cola warehouse, it gave the Dome their first command central.[229] Previously the board had either met at the Dome, at Bill Perk's office, or at that old stand-by, the Italian Village.

Then, the arrival of the first funds from the SAT grant was enough for the board to, finally, move ahead with its restoration work—or, at least, make the announcement. In April 2011, the Foundation hosted, at the Dome, a ribbon-cutting to celebrate the start of the renovation. For the event, Ben Lowder created a new poster that was sold, and Thomas Zung spoke. Along with speeches by Fuller friends and fans, the day included home tours and Dome-related kid crafts.[230]

Also in 2011, the Library of Congress and the Heritage Documentation Programs at the National Park Service inaugurated a competition to recognize the "best single-sheet drawing prepared to the standards of the Historic American Buildings Survey, Historic American Engineering Record, and Historic American Landscapes Survey."[231]

The award was named the Holland Prize after Leicester B. Holland, the Library's first Carnegie Chair of Fine Arts and chief of its Fine Arts Division. In his posts with the Library, Holland launched several significant architectural collections. According to the Library, "The prize is intended to increase awareness, knowledge, and appreciation of historic sites, structures, and landscapes throughout the United States while adding to the permanent HABS, HAER and HALS collection at the Library of Congress."[232]

Thad Heckman expounds:

Somewhere along the many discussions of preservation strategies, a myriad of personalities, various local politics, both pleasant and sundried, I became aware of the Holland Prize. The Dome seemed a perfect fit and, after all, I was right in the middle of doing the "homework" for the Dome preservation anyway, I thought I might as well enter the competition with a carefully-crafted drawing of the house.

A bit old school, I still do much of my design work by hand rendering (which I love), so I decided to take on the Holland Prize single-sheet drawing competition with a good-old fashioned pen-and-ink drawing designed to reflect the drawings from the 1960's, emulating the existing drawings for Bucky's Dymaxion Car (1933) and his other inventions of earlier eras. It did require a bit of a shift in preservation drawing strategies—but not a lot....

Some months passed. Then, in May 2013, I'm in my studio and the telephone's caller ID displayed "LIB OF CON," which I immediately took as a solicitation. I would often get calls from "WASH D.C." and it was always a solicitation. I was sure this was the same and was going to let it ring through to voicemail. I had forgotten about the Holland Prize and, not only that, the submission was to the National Park Service so I didn't make the association with the Library of Congress. After the third ring, knowing voicemail would pick up momentarily, I decided I should pick the call up as this call seemed different. It was.

It was Ford Peatross, Director (retired), of the Center for Architecture, Design and Engineering at the Library of Congress, congratulating me on winning the inaugural Holland Prize. He noted the LoC would be forwarding the appropriate award certification, release form for the image that I am proud to say is now owned

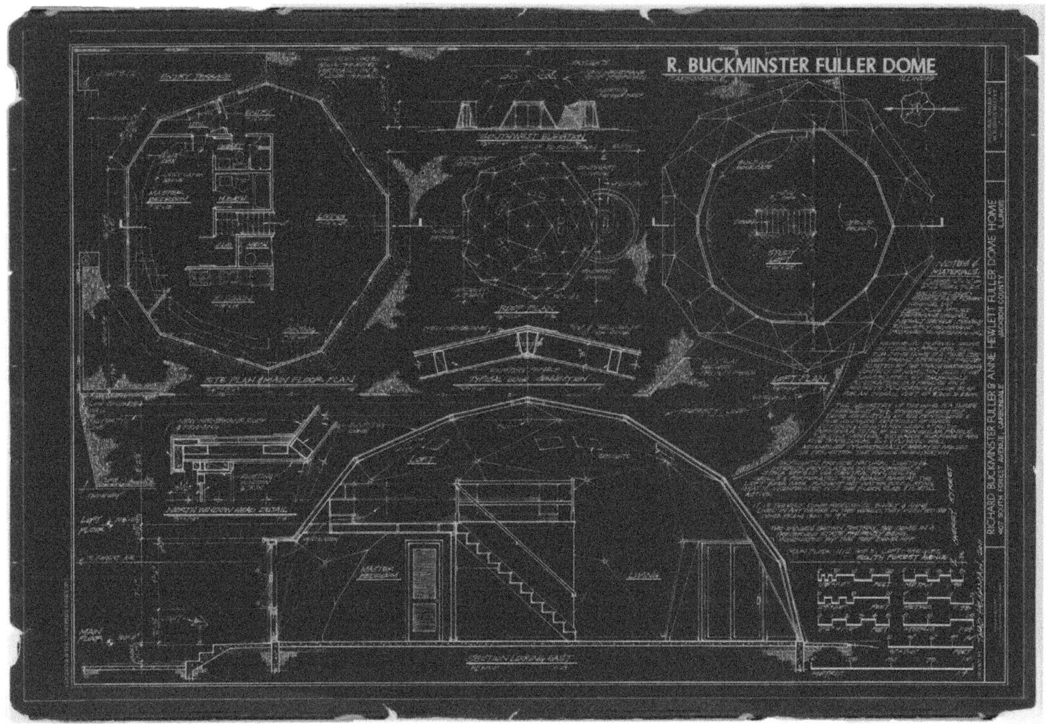

Thad Heckman's award-winning Holland Prize drawing. (*Courtesy: Thad Heckman*)

by the "People of the United States of America" and, of course, the prize money that was donated to help preserve a wonderful little Dome on South Forest Avenue in Carbondale, Illinois.[233]

Although Heckman cannot profit from the Holland drawing personally as it now belongs to the People of America, the NFP can sell prints of it with proceeds going toward the Dome.

With these activities and the influx of some state money, it was predicted the Dome restoration would be completed by 2012. Yet even then, Janet Donoghue, NFP development director, knew that ample work was still ahead for the group. She said at the time, "It's going to take everybody and everything."[234]

Donoghue set up office in the NFP's new space in 2012—just in time to welcome some more new board members. Four new members were added—Terry Hickey, a Carbondale resident and community organizer; banker Stephen Schauwecker; Peter Bahn, a patent attorney hailing from Mt. Vernon (who also made a significant financial contribution to the cause); and Jessica Allee, an assistant to Heckman.

Terry Hickey's known abilities in marketing, soon saw her recruited to the board by her friend Janet Donoghue. During Terry's three-year stay on the board, she worked as the group's treasurer, secretary, and PR guru.[235]

Peter Bahn came to the board via Thad Heckman. The two men were both members of the same group of inventors. Patent attorney Bahn was a boon to the board as it was wise for the NFP to occasionally have counsel within easy reach at this time.[236]

Though her tenure on the board was short (not quite a year), Allee also took on an important role with some of the hyper-detailed work necessary for the home's return to its origins. Allee has a degree in architecture from the Pratt Institute. Once on the board, Allee threw herself into a level of minutiae that, she says, "Was exciting for me."[237]

I got to research historic door fixtures and pursue what type of roofing material and membrane would be an option for the home. I created a spreadsheet that outlined all the pros and cons of each type of membrane: what its cost would be, how to best achieve the look of the original, how sustainable would it be.[238]

Allee also delved into all the home's plywood matters:

Of the home's out-facing flat [vertical] surfaces, only one [of the five trapezoidal shapes] still had its original plywood. The others have been replaced, probably (judging by the material used) some years after the Fullers left. I assume whoever did it, when they needed fixing, they just replaced the entire panel because it was so inexpensive. We could tell which were original by various factors including the space between the vertical grooves, the widths between. So, I did research on whether the original plywood was still being manufactured. Or, if not, where we could find plywood that mirrored the original thickness of the plywood and then, maybe, have routes cut out of it to match what was originally on the house.[239]

… I got quotes and options and turned my findings over to Thad.

Ultimately, when it came to the plywood for the house, a custom order was placed. Though the still-manufactured plywood type T1-11 was close to the original on the Dome, it did not exactly match the old grooves, so something slightly different had to be fashioned by a manufacturer.[240]

The final "newbie" board member for 2012 was Stephen Schauwecker. Schauwecker is an executive VP at Carbondale's First Southern Bank. The Bucky board sought him out for his "money mind." Very active in the community, Schauwecker says today:

I didn't think they could get me on one more local committee … but they did![241]

I was worried, at first, because I wasn't that knowledgeable of Fuller beyond "He designed domes" but I knew it was a worthy cause and important to the community. And that we have a responsibility.[242]

After joining the board, Schauwecker made his maiden voyage into the Dome. He says, "It was a lot more spacious than I thought it would be!"[243]

In June 2012, the *Southern Illinoisan* newspaper noted that the NFP was now only $38,960 short of their SAT matching-fund challenge. They now had until the end of 2012 to obtain the remaining funds. In the article, it stated that once these final dollars were raised, contractors could then be invited to bid on the work of the restoration.[244]

Later in 2012, the NFP got some additional assistance when the city of Carbondale decided to redistribute some of the earnings from its lodging taxes. Of the $120,000 to be divided up, $10,000 was given to the Bucky Dome. The next year, the city continued to support the Dome by making a $10,000 donation to the NFP.[245] These funds were, also, to be applied to the SAT matching funds.

Unfortunately, not all the news of 2012 was good news. A rift between some of the board, which began in 2011, became a chasm in early 2012. At the heart of the matter was Thomas Zung's tome *Buckminster Fuller: Anthology for the New Millennium*. Originally published in 2001, around 2010 or 2011, Zung approached the NFP about funding a second printing.[246]

Some board members leapt at the chance to reissue the book, believing that assisting in the spreading of Bucky's name would have a positive trickledown effect for the restoration. They further believed that the book would make a handy premium—good giveaways to donors.[247] However, other board members—including Brent Ritzel—bristled at the hefty price tag that was needed to reissue the book: $15,000. Ritzel says today, "That was actually about the same amount of money we needed to finish the SAT matching funds! But, instead of applying that money from our budget to that, they wanted to have copies of this book!"[248] Despite his objections, the board eventually paid for the book to be reissued.

That disagreement begot others. Ritzel, who had been heading the board since 2009, slowly saw his frustration with some aspects of the board begin to outweigh the good he thought he could do. By this time, Ritzel had already announced his candidacy for mayor of Carbondale and was headlong into his campaign when he decided to resign from the board in 2012.[249]

The various bumps in the road experienced by the board in early 2012 did not fully recede as the year wore on. The NFP received a devastating blow in August 2012 when one of their founding board members, John R. Johnson, was killed when his vintage private plane, a 1956 Piper Apache, in which he was flying, crashed into a pasture near Canton, Missouri.[250] To this day, it remains unclear who was piloting the plane; Johnson was aboard the craft with his friend and fellow pilot Carl S. Maiden. The duo had left Pinckeyville on the morning of August 29 *en route* to an antique aircraft show in Iowa when the plane went down. Both men were killed.[251]

The loss was shocking to the group. Mr. Johnson was highly accomplished and he kindly called upon all his skills with computers, engineering and tinkering to aid the board over the years.[252] Johnson was survived by his wife, daughters, and stepchildren. The family graciously listed the Dome as one of the charities that memorials could be made to.[253]

Despite the tragedy, the NFP could, at least, be fortified with the steady trickle of funds coming in. Yet would it be enough to meet the SAT deadline looming at the end of the year? Over the years, monies for the Dome have been obtained from generous—and often anonymous—donors, some giving amounts of five figures. Additionally, the American Institute of Architects once kicked in $1,200.[254] Meanwhile, donations of materials and labor has come in thanks to generous locals like Tim Mitchell and his company, Universal Glass. Mitchell has regularly accommodated the home's needs without invoicing a cent.[255]

After the difficulties of 2012, 2013, thankfully, kicked off on a good note. In January, $1,500 was gifted to the Dome from Landmarks Illinois's Preservation Heritage Fund.

The late John Johnson, doing what he loved best—flying. (*Courtesy: Sabrina Hardenbaugh*)

That money was added to the matching funds requirement.[256] These final sums, and the completion of the bidding process, meant that restorative work was eminent. In early January, the NFP issued a press release announcing that work would start in the spring.[257]

In April 2013, SIUC's Morris Library remounted an earlier exhibit devoted to Buckminster Fuller. *Inventions: Twelve Around One*, originally staged in 1981, featured Fuller designs and structures. The featured items, all donated by Bucky's daughter Allegra, included ten 40-inch × 10-inch drawings by Fuller. The exhibit's opening night reception featured remarks by new NFP President Jon Davey and former NFP board member Larry Busch.[258]

Jon Davey first encountered Buckminster Fuller when he was in the military in Germany in the early 1970s. He used to visit a coffee shop on the base, and there, on the wall, was a poster with a Bucky quote on it—"God is a verb, not a noun, proper or improper." Later, in 1976, Davey saw Fuller himself when Bucky spoke at the SIUC's Shryock Auditorium.[259]

A native of Buffalo, New York, Davey lived a childhood made up of many moves. He says, "I attended six different grade schools as we moved around the country." After military service and attending Rend Lake College, Davey attended SIUC,

Jon Davey.

graduating in 1979. He joined SIUC's faculty in 1981.[260] Now a professor in the School of Architecture, Davey has been active with the Dome for many years and became president in February 2012.

It was under Davey's guidance that the first ever Dome Days took place in April 2013. The three-day festival (April 18, 19, and 20— timed to coincide with the month in which the home was originally built) included the opening of the Dome for a reception and the display of various Bucky artifacts as well as kid activities. Over on SIUC campus, also as part of Dome Days, a session of the World Game was staged.[261]

Those inaugural Dome Days included a performance by Mel Goot and his local jazz trio: Coulter, Goot & Wall. After that performance, Goot's trio became a Dome Days tradition, they would also play in 2012, 2013, and 2014.[262] Also in 2013, in order to take advantage of the geodesic's legendary sound properties, then NFP board member Jay Needham (joined 2012) reached out to conceptual musician Stephan Moore to stage a sound installation in the home as part of a Dome fund-raising event.[263]

Musician Stephan Moore pictured during his sound installation. (*Courtesy: S. Moore/Jay Needham*)

Moore's work, called *A Better Place*, was a sound installation performance piece that posits a fictional conversation between two of the twentieth century's better-known aspirational thinkers—John Cage and R. Buckminster Fuller.[264] *A Better Place* debuted at the Dome on April 19, 2013, and stayed in place through June.[265]

In 2013, thanks to all fundraising and hard work, tangible proof of progress was now on the horizon for the Bucky Dome. Though the group still needed to meet their matching-fund requirement, they were able to submit their first "draw" on the SAT grant to collect some of the promised funds. This allowed the group in July 2013 to open the bidding process to contractors, letting them vie for the job of restoring the Dome's exterior.[266]

To solicit bids (again, per SAT guidelines), the NFP published an invitation to bid in area newspapers and in online "Plan Rooms" frequented by architectural firms and contractors. To get the best selection, the board also contacted directly various dome-oriented businesses to submit their applications.[267]

After the bidding was completed, three firms had put in applications. However, with a bid that came in a full $100,000 less than the second nearest bid, Blair Wolfram of Domes, Inc. was awarded the contract on August 16, 2013.[268] Once the contract was awarded, a preliminary schedule was drawn up. Work on the Dome was set to begin in the spring of 2014.[269]

Never one to let an occasion go by unobserved, on April 5, 2014, the NFP held a special "ground-breaking" ceremony at the Dome.[270] A few days after the ground-breaking, "Dome Guy" Wolfram packed up a diesel truck with a 30-foot trailer behind it, loaded it with tools, and drove from his home in Minnesota. With his right-hand man, John

Kennedy, also in tow, the two rented houses near the Dome to get ready to start their work. They would be on-site for the next five months.[271]

Monday, April 21, 2014, is an important date in Bucky Dome history. That morning, a work crew put up a safety-oriented construction fence made of metal posts, chain link, and black scrim around the edge of the Dome property. Then work on the Dome finally began. The first order of business was to tear off the layers of shingles that had for so long covered the 1,675-sq. foot roof.[272]

The removal of the shingles brought with it a bit of a surprise. Discovered underneath some of the asphalt was a second shade of blue paint. Previously unknown of except when the Dome was built in 1960, the home had two different hues of blue color: one shade of blue on the ten vertical sections and then a lighter shade—meant to emulate sky or the horizon?—on each of the five "canopies" around the door frames. Long covered up, this unique design element had been forgotten. Today, the home's original colors have been restored.[273]

Once the outer shell of shingles was skimmed, the real work of the roof repair could begin. Rachel Stout, in 2015 article for a construction trade publication, recounted the Dome's re-roofing process:

First, all of the wood sheathing had to be inspected and the insulation joints had to be sanded in order to achieve the desired aesthetics. Next, each of the sixty triangular pieces received a new layer of insulation bevel-cut to match up with the adjacent pieces of insulation. The insulation was adhered to the plywood roof deck. Because of the various slopes on the geodesic roof, the insulation was temporarily fastened with screws and insulation plates to allow the adhesive to cure. Once the adhesive was cured and the insulation was solidly adhered to the roof deck, the temporary fasteners and plates were removed and the new synthetic membrane roof or TPO (Thermoplastic Polyolefin) membrane was cut into sixty matching triangular pieces and fully adhered to the insulation with solvent-based bonding adhesive. At each splice, caulk lines were marked to ensure that the TPO flashing overlapping the splices was set perfectly straight, with no more than 1/16" change in straightness.

Once the TPO membrane was installed, the rooftop was pressure-washed to test the water-tightness of the system and to restore the membrane to its original whiteness. Two custom-color acrylic coatings were then applied to the membrane to replicate the original colors of the Dome Home rooftop….

The rooftop restoration … was completed in less than a month by Allstar Roofing of Maple Plain, Minnesota. Covered by Carlisle's 20-year Total System Warranty, the new rooftop will provide the Dome Home with much-needed long-term protection from the elements while helping to restore its original owners' vision of structural integrity and picturesque design.[274]

With the roof renewed, in the summer of 2014, the internal support pole could (finally) be taken out and the long-standing over-dome was ready to be removed as well. Originally, the removal date was July 8, 2014; that was when the team of roofers were scheduled to arrive on the site. Yet once they got started on the roof, the workers found the over-dome frame to be handy for their ongoing work. Therefore, the outer dome was not actually removed until about a month later.[275]

Right: Contractor Blair Wolfram pictured in the walkway space between the home and the over-dome.

Below: Once the shingles were removed from the home, Bill Perk (pictured) and others were delighted to see that the blue of the Dome's original exterior was still intact.

Jon Davey and a worker survey the roof as its shingles are finally discarded.

After the over-dome was removed, in a lack of waste that would have delighted Fuller, it was not destroyed but reutilized; it was relocated, becoming part of Carbondale's Gaia House Interfaith Center's newly created Labyrinth Garden. It stands there today.[276]

True to form, the removal of the over-dome was another multi-member board endeavor. One Saturday in October 2014, a small group of board members, including Jon Davey, showed up and began pulling off the protective tarp and deconstructing the shell itself.[277]

One of those volunteers that morning was future board member Peter Szczesniweski. Szczesniweski was a junior at SIU and was in one of Davey's classes when Davey announced they needed help at the Dome over the upcoming weekend. As Szczesniweski recalls, "I remember thinking 'That's awesome! We have homework!' I volunteered. I think I was the only student on that first day of work."[278]

A native of Glen Ellyn, Szczesniweski was somewhat aware of Fuller in Carbondale when he first moved to town. He got lessons in all things Bucky soon after, however, when he moved into a house only a few doors down from the Dome.[279]

Szczesniweski's assistance with the over-dome was also his first trip into the Dome. He recalls, "I was most struck by how open the inside feels. From the outside, it looks small and cramped, but it's not. It shocks me at the volume."[280] Shortly after that day of work, Peter inquired about joining the board. Szczesniweski attended a meeting shortly thereafter and was voted in.[281]

The visible changes in the Dome did not go unnoticed. The *Southern Illinoisan* announced the start of the roof renovation with a 2014 story that reported the good news. The start of its headline was "Bucky Dome restoration project lifts off" and its end echoed the feelings of many by adding "finally."[282]

Right: The Dome as construction site. Note the fully rebuilt "hex" at the top of the photo.

Below: The roof of the Dome completely stripped of its shingles.

However, for all the progress the Dome was making, they were facing a new concern: donor fatigue. For ten years, people had been reading about the Bucky Dome, celebrating Dome Days, and donating money. Unfortunately, much of the progress on the Dome had for too long been hidden under the home's over-dome or been absorbed by inflation.

When the home was first acquired in 2002, it was estimated that a full restoration would cost about $100,000.[283] By 2005, after further inspection, that amount was amended to $240,000. Two years later, the amount was $265,000.[284] A decade later, it was edging over the $300,000 mark.[285]

As time wore on, many began to wonder just when the Dome was going to be finished. It was not just the general populace that could grow weary; sometimes, so did board members. Board member Janet Donoghue eventually felt it necessary for her to step aside. "I realized," she says, "That the NFP is, in many ways, a relay race. We do what we can and then move on and let a new generation take over."[286]

Some of that new energy came via board member Jay Needham. Originally from Chicago, Needham earned his BA in 1986 at SIUC in the field of cinema and photography. He was back in Carbondale, *c.* 2004, after earning an MFA at the California Institute of the Arts, and was now teaching at SIUC's Department of Radio, Television and Digital Media. One night, he was walking down that same old Carbondale street when he spotted a light from inside the Dome. He went closer and inside could see a small group gathered around a card table. A NFP meeting was underway; Needham crashed it.[287]

Later, after formerly joining, Needham found synergy with his "day job." He reworked his department's internship program to allow students to become involved with non-profits, giving them professional experience with it and the Bucky organization various new soldiers.[288]

After the over-dome's removal, for the Bucky Dome, it was full speed ahead. First, they got good news regarding their SAT Grant—they had met their matching-fund requirements. Crunching the numbers, the board submitted their final totals to SAT authorities on June 23, 2014.[289]

In July 2014, the *Southern Illinoisan* announced that Phase I of the restoration had now reached its halfway point. Interviewed on the job site, Blair Wolfram noted that although he and his crew were about ten days behind schedule, progress was being made.[290]

At the time of the interview, Wolfram had just finished with the roof. Today, the roof of the Bucky Dome has three layers: there is new plywood sheathing over which are sheets of recovery board and over that is a layer of Carlisle TPO (thermoplastic polyolefin) membrane. The membrane will provide the roof with long-term durability and flexibility. This outer shell should, like most other roofs, endure for about twenty years. Meanwhile, underneath all three items still rests the home's original 1960 plywood; it has been left in place in case a day should arrive when someone wants to see or study the home's earliest incarnation.[291]

It was Thad Heckman who, after considerable research, derived the roof covering's final "formula:"

> I agonized about what was the BEST roofing system to put on the Dome that would offer substantial protection for years but also emulate as closely as possible the original appearance of the Dome…. I presented the final options to the board. The NFP made the final decision.[292]

The newly recovered roof.

The Bucky Dome with its new, gleaming, restored and weathertight roof.

Also underneath the membrane *et al.*, but over the plywood is an outer metal "tension ring," a continuous strap of metal, welded at the ends, and affixed to the roof and walls of the Dome that completely encircles the house. This was a Blair addition and departs from the home's original architecture but is a necessity; in the coming years it will keep the Dome from ever losing shape again.[293]

With the roof finished, Wolfram *et al.*, moved onto the skylights. He said, "Restoring the ten skylight windows posted a challenge as no one within the industry made the particular size we needed."[294]

One of the Dome's most distinctive features, the skylights, had long been problematic. Originally, Bucky installed ten of them, arranged in a circle near the very apex of the Dome. They let in light and gave an indication of the sky above. They were flush with the roof, and they leaked. So, Bucky went back to the drawing board. At some point, probably after the first shingling, the skylights were covered with a series of five larger skylights (measuring roughly 3 feet by 8 feet). Each of these newly added bigger skylights covered two old skylights. Hence, the number of skylights visible from the outside was reduced by half. In the vernacular of the building trades, each of these new skylights were also "curbed," or set up to stand higher than the roof. This is done with most skylights today as it assists with weather proofing. Yet, if we believe many of the Dome renters, even these curbed skylights often caused problems. At one point, someone simply covered over the skylights completely. These obliterated any light coming in but, also halted (most of) the water.[295] Today, the skylights have been restored to their original number and are again flush with the roof.

With the outside of the roof now restored to its original glory of ocean blue(s) and pristine white, Blair's workers were now free to approach other aspects of the Dome's exterior. Though Blair Wolfram acted as the contractor for the restoration, he subcontracted to outside trades for other items. Chuck Grammer of Chuck Grammer Construction, in the neighboring Carbondale town of Murphysboro, was one of those Wolfram brought onto the project. Though Grammer says he "always knew the dome was there," he did not have much knowledge of Buckminster Fuller before being hired. After being retained, he quickly turned to the web to study up.[296]

Grammer was at the Dome, off and on, for three months. Assisted by Bill Quigley, the two men took on the task of reframing the doors, windows, and geodesic frame where it was needed. Rotted wood, bad repairs, and wear and tear necessitated the removal of every window and door. Each item was then packed up and transported to Tom Guelcher of The Turning Point Woodworks, Inc. in St. Paul, Minnesota.[297] Guelcher's company specializes in tough woodworking jobs and historical restoration. He was discovered by Wolfram via a website devoted to historical restoration.[298]

Once the four windows were excised from the home, Wolfram loaded them onto his truck and hauled them back to Turning Point. Guelcher spent about three months repairing the four windows. He estimates that he was able to salvage about 75 percent of the original windows' wood and framing. The original glass from all but one of the windows was reused. One windowpane, however, was just too far gone. For it, Guelcher acquired another pane dating from 1960.[299] At the end of the summer, once Guelcher had completed his work, Wolfram once again loaded up the windows and returned them to Carbondale.

Above: Though the door seen here was still awaiting replacement, the front of the home was now looking more like it did when the Fullers lived in the Dome.

Right: The Dome's replaced front door viewed from within.

Back on the site, the windows were reinstalled and the framing around them was reconstructed to be level and plumb. The craftsmen then reinstalled every single original window and sliding door. The home's primary "front" door, however, was deemed too damaged and had to be fully replaced; this was not a crucial loss as the then front door was not the Dome's original anyway. For the new front door, and following NPS guidelines, the board ordered up a custom-made replica of the home's original.[300]

On site, for Grammer and Quigley, their work was methodical and decidedly uncomfortable. With the protective dome still in place during most of their time there, and with little ventilation, even a pleasant day could turn the Dome's interior into a sauna. One afternoon, Quigley brought in a thermometer and measured the temperature—114 degrees.[301]

Calling his Dome work a "unique experience," Grammer also attended to the Dome's roof, plywood pieces, and overall symmetry: "This was my first dome but I could tell it was out of shape. Especially the south part of the dome. There was a big pole holding up one of the vertexes. We had to move everything around to see how much we could salvage and repair."[302]

As the roof had leaked, it caused not only the plywood triangular faces to rot but the original 2 × 3s that formed the edges of each triangle to decay. Though some supports were too destroyed to be salvaged, others could be saved with a little carpentry work. That involved the addition of new and matching 2 × 3s secured to each side of the original strut, reinforcing it.[303]

Though Grammer and his team found some parts of the home beyond redemption, other aspects were in okay shape. Grammer recalls, "I was relieved that the [upstairs] bookcase could be left intact."[304] Grammer adds, "Luckily, Blair and Thad had all the numbers" for the dimensions of the Dome.[305]

Also brought into the project was All-Star Construction, located in Minnesota. Today, Braden Larson, who served as project manager for All-Star, admits he was only barely familiar with Buckminster Fuller. However, he learned quick after he was hired. His job was to put on the final covering of the roof.[306]

Though Larson never visited the Dome in person, he did have to figure out a work plan, buy the membrane, tabulate the mathematics, and put together a three-man crew to dispatch to Southern Illinois.[307]

At the worksite, the All-Star workmen commenced with the tedious task of seaming all the roof-top triangles, sanding the joints, and tightening each to within one-sixteenth of an inch.[308] Then, they put on the sheeting of the membrane. Originally expected to take one week to complete, the job expanded to three times that. Further complicating the work was the difficulty of maneuvering around the roof's slope, as it was far more precarious than working on a normal, slanted rooftop. The curvature of the Dome necessitated that each worker wear safety equipment prior to going up.[309]

Roof work finished on October 19, 2014.[310] Its completion revealed the new, glistening outside of the Dome, and it was cause for celebration. A special ribbon-cutting ceremony was held on the property on September 17, 2014. Jon Davey presided over the event; those in attendance included Allegra Fuller Snyder (who flew in from in California), Blair Wolfram, Thomas Zung, Bill Perk, and other NFP members.[311]

The Dome's "polar-pent," or uppermost tip, was almost completely disintegrated by the time the restorationists began their work.

The rebuilt polar-pent.

Along with announcing the conclusion of Phase I, the ceremony officially kicked off Phase II, which would focus on the home's interior, including rewiring the electrical, installing insulation, recovering the walls, restoring the cork flooring, and painting. At the marking of the end of Phase I, Bill Perk stated, "We have to raise another $125,000 to $150,000 to return the interior to how it's supposed to be…. It's not a matter of years or manpower, but money."[312]

Davey stated it was the hope of the NFP that not only would the necessary monies be raised but that Phase II would be completed by the fall of 2015.[313] By the time of the announcement, Thad Heckman had already begun work on the necessary drawings for Phase II.[314]

As Phase I concluded, some additional attention got showered on the Dome. In January 2014, Baysinger Architects of Marion issued a list of the "Top 10 Architectural Wonders of Southern Illinois." Carbondale's Bucky Dome was included alongside SIUC's own Brutalist style Faner Hall; the 5th District Appellate Court House in Mt. Vernon; and the Historic Opera House in Sesser.[315]

Coinciding with the end of Phase I, Allegra returned to Carbondale to take part in the discussion "Our Sustainable Legacy & Future: Buckminster Fuller at SIU." Held at the Guyon Auditorium at SIUC's Morris Library, Mrs. Snyder was joined by playwright D. W. Jacobs, who had previously performed his one-man show, *R. Buckminster Fuller: The History (and Mystery) of the Universe,* at Morris, and Benjamin Lowder as well as NFP alumni Brent Ritzel and Janet Donoghue.[316]

Yet if 2014 was good, 2015 was even better. First, new board members arrived. For Daniel Presley, joining the board seemed inevitable; he grew up about two blocks from the Dome, on Oakland Avenue. Though originating from Carbondale, Presley departed at age seventeen to study at Eastern Illinois University. After earning his BA in 1994, Presley later pursued an MA in Business Administration at Pepperdine; he graduated in 2005. After that, Presley remained in LA for several years. He returned to Carbondale—"for good"—in 2012.[317]

Today, Presley works for Bank of America's Los Angeles office but telecommutes and continues to live in Carbondale.[318] Presley has also come full circle. Today, he lives in a house in the Arbor District, not far from the Bucky Dome.[319] Since coming to the board, Presley has assisted with everything from electronic marketing to cleaning up the grounds after Dymaxion Days.[320]

Also joining was Lindy Loyd, who grew up in the Chicago area but has lived in Carbondale for the past forty years: "I came to Carbondale in the summer of 1975. I was hitchhiking across the country. I was on my way to California to be the next Joni Mitchell but I stopped over in Carbondale to see my sister and never left."[321] Once settled, Loyd went to work for Carbondale High, where she was its AV technician for thirty years until retiring.[322]

It was during her time at the high school that Loyd became alarmed about how few people seemed to know of the Dome.[323] To renew attention on the Dome, Loyd began to encourage students to explore the topic of the Dome and domes in general for the annual school science fair. Soon, Bucky was a popular subject among the kids, and at least once, a project based upon the Dome made it to the science fair finals in Springfield.[324]

At the unveiling of the Dome's restored exterior, Bill Perk (far right) is joined by (from left) Thomas Zung, Allegra Fuller, and Thad Heckman.

Interior of the bedroom in the Bucky Dome.

Shot of the east side of the upstairs loft; its original wraparound bookshelf is still in place.

After retiring, and at the behest of her friend Judy Ashby ("Judy kept pushing me," Loyd says), Loyd joined up with the board. Since then, her duties have run from helping with the annual Dymaxion Days to the weekly mowing of the Dome's lawn during the summer.[325]

In 2015, the NFP got a vote of confidence when, in May, Carbondale cited the Dome as that year's recipient of its Historic Preservation Award. This honor is bestowed to the local person or people who have done the most during the past twelve months to restore an area landmark.[326]

Then, in August 2015, the NFP was informed by Landmarks Illinois that they were that year's recipient of the Richard H. Driehuas Foundation Preservation Award for Restoration. This honor recognizes "innovative preservation projects, dedicated organizations and uplifting success stories from across the State of Illinois." The award came with $500 in prize money. A reception was held on October 17, 2015, in Chicago. Jon Davey attended along with Bill Perk and Thad Heckman.[327]

Only a few days after the Chicago reception, the Dome was the site of a major photo opportunity when Nashville's Lane Motor Museum shipped to Carbondale a replica of Fuller's Dymaxion car. The meeting of these two Fuller icons was set in motion by Thad Heckman when he placed a call to the museum after he learned of the existence of the car from friend and university colleague, architect Peter Smith.[328] Heckman noted:

I knew that along with Lord Foster, Jeff Lane of the Lane Museum also had a rebuilt Dymaxion. So I decided to call Jeff Lane and, after a brief introduction, I asked if the Museum would consider bringing up their replica of the Dymaxion.

I commented tongue-in-cheek that it would be much easier for him to bring his Dymaxion car to Carbondale rather than to bring the RBF Dome to Nashville. I also noted it would be the first time ever in history that a Dymaxion Car had been next to the Bucky Dome—two of the last century's great design artifacts together at the same time.[329]

Lane accepted immediately.[330]

Lane's auto museum is a not-for-profit established in 2003 and was originally founded to feature his family's personal collection of seventy vehicles of mainly foreign-built cars. Today, it holds over 450 vehicles. They are especially fond of vehicles that are three-wheeled (like the Dymaxion), amphibious, or which operate on alternative fuels.[331] Having studied the Dymaxion plans for a long time, Jeff Lane eventually decided to build a replica to see if the car was everything that Bucky said it was.[332]

It took Lane eight years to manufacture the Dymaxion replica. Though the exacting replica of the car got a thumbs-down from *Autoweek* (who faulted the vehicle's steering and stability), the Dymaxion has proved popular.[333]

On October 23, 2015, the Dymaxion vehicle was brought to Carbondale to be part of Carbondale's first Dymaxion Days, now an almost yearly event. Along with the chance for Fuller fans to grab a photo of these two great icons side by side, free rides in the car were also offered. The day concluded with a chance for everyone interested to make an examination of the reproduction.[334]

In the end, over 150 people turned out to see the car. Even NFP board members found themselves a little awestruck at the occasion. Though Judy Ashby did not get a ride in the vehicle, she says she still had fun "seeing it, getting to touch it ... fondling it!"[335]

The visit by the Dymaxion Car yielded over $2,000 for the NFP.[336] Still, despite this money—and the generosity of various donors—coffers for the Dome were short by the end of the year, causing the NFP to retract their prediction that Phase II would be done by the close of 2015.[337]

Despite that projected shortfall, the Bucky Dome in 2015, and early 2016, continued to be a hive of activity, and though the architectural drawings for Phase II and Phase III were still a work-in-progress, some preliminary work concerning the interior got underway.

First up was the home's flooring. Originally, the Fullers lined their floors with cushiony cork. Over the years, wear and tear as well as water damage demanded that much of this flooring be removed or be painted over. Yet, the NFP wanted to get the floor again back, close to the original. One day, Thad Heckman realized that the original cork flooring extended all the way into the Dome's bedroom closet, and those floors had never been damaged, pulled up, or hardly even stepped upon. Additionally, since it was "buried" within the house, that cork had not even been exposed to sunlight. Getting down on his hands and knees, Heckman crawled back into the rear space of the closet and carefully took up two tiles to use them for color, type, and thickness matching.[338]

Researching cork providers is probably not everyone's idea of a good time, but it is what Heckman had to do. After learning that the original cork vendor, dating from 1959, was no longer in business, Heckman headed to the internet.[339]

Two great Fuller icons are seen in one photo for the first time as a Dymaxion Car is posed outside the Dome in 2015.

Crowds gather at the Dome for the first visit of the Dymaxion Car replica.

First, it was determined that the cork in the home was of two different thicknesses. On the main floor, it was about 5/16 of an inch thick, but the cork in the loft area was much thinner, only about 3/16 of an inch. The upper and lower levels also have different colors—upstairs, it was a light, sandy brown; below, it was dark, almost a chocolate mottled brown.[340] After the necessary web surfing, Heckman found one or two companies able to match the original cork.[341]

However, with the Dome being a historic property, it is not as simple as calling them up and placing an order.[342]

First, for any possible purchase made out of "public money" (i.e. money from public grants, etc., such as the SAT Grant), the board first has to send out for competitive pricing. The use of grant money requires the board to first identify at least three companies offering roughly the exact same product. Once bids are received from these three companies, the board is then compelled to go with the company offering the lowest price.

That is, unless the product the NFP is seeking is "single sourced," meaning only one particular company has what the board is seeking. If a cork company, for example, is the only business that offers the product that the board needs, then the board—after filing the justifying paperwork—can purchase directly from them. In contrast to all this, however, if private money is used, the board can simply select products and contractors according to their own wishes.

Around the time of the great cork investigation—late 2015—Thad Heckman brought into the home Rick Cantwell of REC Management, Inc. for the home's next evaluations. Cantwell—who donated his services—is a "moisture guy" and had the task of determining the home's moisture content as well as its antimicrobial application and fire mitigation, all steps necessary before the installation of new wall insulation.[343]

To the relief of all, Cantwell found that the Dome was in very good shape, far below government-recommended standards. In the meantime, a dehumidifier was purchased to mitigate interior moisture until the proper insulation could be afforded. The Dome has been under constant dehumidification for quite some time—years in fact, after the review by REC. However, it was not until much later, in spring 2018, when Ed Cook was getting ready to start Phase II, that an ozone generator was placed for four days to remove any microbial or fungal activity—it if existed.[344]

Cantwell's finished assessment meant that the NFP was now free to continue its fundraising for the purchase and installing of home insulation; that cost was estimated to be around $9,000.[345] The job went to bid in late 2016.[346]

However, for all the good news about moisture, the board then had to reface an old foe—the skylights. Though they now looked as they did the day Bucky installed them in 1960, they also had the same problem—too much moisture.

The rebuilt skylights consist of two stacked panes of glass with an airspace between them and a third plastic protection layer over that. Between the panes, condensation was developing. The moisture generated, thankfully, will not be absorbed by the new roof membrane, as the membrane can resist that moisture for decades to come, but it is still not an ideal situation. Therefore, the team went back to the drawing board, but this time with a custom designed truly insulated (double pane and sealed) glass isolating the skylights from the adjacent wood framing. Thad Heckman noted that this should have been done the first time around. "This time," he says, "We will not compromise on an originally questionable design."[347]

Photo showing the contrast in appearance between the home's original flooring (upper part of the photo, taken in the bedroom closet) and the bedroom's flooring as it looked at the start of the restoration.

One of the newly installed skylights.

While the skylights were being resolved, the board then chose to tackle another old enemy: the fence. After years of make-do patches, by the early 2010s, as the home's exterior and roof were repaired, only a few portions of the original fence were still standing. The remaining original sections were on the east side of the driveway and on the north side of the home. With regard to the latter, it appears only the fence posts are original.[348]

As mentioned, most of the original fence had fallen down. These pieces have been systemically gathered, labeled and stored by the board until the day comes to attempt the fence's resurrection.[349] During some unseasonably warm days in February 2016, preliminary work on fence reconstruction began.[350]

Thanks to the still-standing portion, a few post stubs, and vintage photographs (some from former Dome owner Mickey Mitchell), Thad Heckman and Larry Weatherford could begin to estimate the exact perimeters of the fence as Bucky had originally placed it. However, to get the exact location of the long-ago abandoned southeastern corner post, the two men relied on something far more basic: namely stabbing into the ground with a long screwdriver, digging down, and hoping to find something to evidence the fence's original corner placement.[351] They found it. Heckman explains, after hitting something:

> I dug down and found chunks of hardened wood that turned out to be the base of the redwood corner post—and it was exactly where we thought it should be, so, I think, we have the location now—or at least VERY close. We can now show the fence in the Phase III drawings in the proper locations with the proper post spacings.[352]

With the property's temporary construction fence visible in the back, the mini-dome covers the front-yard fountain.

With renewed attention being paid to the Dome grounds, it seemed natural to also begin to focus on the front yard's distinctive fountain. Years ago, someone had stolen the motor from the fountain, and then the elements took their toll on the pipes and concrete that completed the garden feature. When the NFP assumed control of the house, they put the fountain on the back burner and usually left it covered with plywood or placed their small metal geodome over it. Then during those same days in 2016, when Heckman was on the property, he got curious. Uncovering the fountain, he found its control valve and turned it. To his surprise, water flowed out. Perhaps the fountain was not in as bad a shape as they all thought.[353]

Granted, water was not emerging from the actual fountain spigot, but water was running into the fountain's pump pit and this was good news indicating that the underground pipes were still intact and operational.[354]

Building codes for yard water features are more complicated today than they were in Bucky's day, so bringing the fountain back would require upgrades. The upside of this is that the company the NFP turned to help with the fountain agreed to donate their services. Asaturian, Eaton & Associates, of Carbondale, are engineers and consultants.[355]

The generosity of Asaturian, Eaton & Associates, and its professional surveyor, Marty Merrill, did not end with just their fountain "know-how," however. AE&A also kindly offered to donate their services in terms of surveying the property and writing the description from their findings. They also offered their assistance with the home's MEP needs—mechanical, electrical, and plumbing engineering.[356] Once Asaturian finished their work, fountain resurrection will be turned over to RP Coatings, Inc.[357] As its name suggests, RP Coatings, owned by Randy Penrod, is a Troy, IL-based company that specializes in producing and applying protective coatings and creating moisture barriers for water treatment plants and large scale industrial and commercial projects.[358]

Working with Thad Heckman on the design of their new warehouse, the two men were at a working lunch one day when Heckman brought up the Dome and the fountain. He asked Penrod what he should use to coat and seal the fountain in conjunction with the new plumbing. Penrod then volunteered his company to bring the fountain back to life. Heckman says, "I didn't even have a chance to ask—and, of course, I was going to ask!"[359]

For the NFP, early 2016 was dominated by the completing of some very important paperwork. A requirement of the SAT grant was the submitting of an "easement," a document that certifies the SAT grant was applied toward a property "in the public interest" and not for personal gain. The easement was completed and forwarded to the National Park Service in March 2016.[360]

Autumn 2016 witnessed the second visit of the Lane Museum's Dymaxion Car. This second visited attracted even more people than it did the first time around. Arriving in town on October 21, 2016, the Dymaxion auto, along with visiting the Dome, was also part of SIUC's homecoming parade. Although he was offered the chance to ride in the car for the parade, the university's then chancellor declined the opportunity in favor of something a little more traditional. Nevertheless, the Dymaxion was one of the highlights of the parade and even attracted the attention of KFVS News from Cape Girardeau, Missouri. They sent over a reporter. The return of the Dymaxion brought in $500 for the NFP's coffers.[361]

Then, Halloween 2016 brought a fresh batch of carved, architecturally inclined pumpkins to the Dome for their annual outdoor display. The pumpkin display would reoccur in 2017, 2018, and 2019.[362]

Throughout its ongoing revival, the Bucky Dome has been a popular local attraction. Even in its pre-restored state, it annually attracted over 100 visitors a month—via tours and school and college groups. In 2016, the NFP began working towards the day when the Dome would be open every day for people to come and see. To that end, board members reached out to local tourism bureaus and hotels to be sure to share the presence of the Dome in Carbondale with their visitors. The NFP reached out to the city, too, for the city to add it to all its tourism information.[363]

Also transpiring in 2016 were an application to the city to become exempt from property taxes (that thanks largely to the efforts of Steve Schauwecker) and the return of an "old" board member—Ed Cook:

[Originally,] I resigned to prevent the appearance of a conflict of interest if I was doing the entire restoration. Turned out my small residential business was not equipped to meet federally-funded requirements for payroll, etc.

Thad approached me about returning and doing phase two of the restoration because the funding is structured differently. I proposed to do the work for my cost. Bill Perk and the rest of the board seemed pleased with my offer and welcomed me back.[364]

Cook was voted back onto the board in February 2017.[365]

At the very end of 2016 and beginning of 2017, despite a rather light winter, the Dome's exterior showed a patina of dirt which dulled its once renewed, gleaming white surface. It was decided that the Dome needed a good scrubbing. The substance to be used was Dawn dishwashing detergent. A proven grime cutter, but safe, Dawn was recommended by a local roofing installer.[366]

Much of the grime deposited on the Dome's roof was found to be from the sap dripping from one of the large trees near the house. A towering cedar now, but only a few feet tall when the Fullers lived there, the board is considering eventually (and carefully) having the tree removed from the lot.[367]

It was also noted at this time that some of the house's exterior paint had not worn well and necessitated another application of one of the blue hues. Further, after thinking they were finally done, Thad Heckman and some of the other board members noticed some issues—again—with the skylights. Water was not dripping into the Dome, but it was accumulating and condensing within the individual panes of some of skylights; this (again) was not what was supposed to happen.[368]

Other bits of business that year included discussion of an upcoming visit by a film crew from the State of Illinois for a "Landmarks Illinois" short film; preliminary plans for that year's upcoming Dome Days; the development of a thorough database for all donors to the Dome; and a complete overhaul of the organization's website, the latter mainly handled by Ben Lowder.[369]

The year 2017 included the observation of what would have been Fuller's 122nd birthday in July of that year at Carbondale's kid-oriented Science Center located inside the town's University Mall. The Center unveiled a new Bucky-focused section. The unveiling ceremony was free, and first-day visitors got to snack on a Dymaxion cake and geodesic cookies.[370]

Later, at the Dome, on December 1 and 2, the house was home to a multimedia exhibition of video, music, art, and performance in what many hoped would become an annual event. This program, titled "Came to Pass, Not to Stay," was a showcase for MFA graduate work inspired by the Dome, Buckminster Fuller, and his universe of ideas.[371]

Despite all this activity, it was what happened in between these events in Carbondale that generated the most attention. In August 2017, the first total eclipse of the sun in thirty-eight years was scheduled to happen, and coincidentally, it would reach its point of greatest duration (totality) while directly over the city of Carbondale.[372]

For well over a year prior to the event, the city worked to make the most of its astrological good luck by turning itself into eclipse central and opening its metaphorical city doors to all visitors. Months in advance, all area hotels and motels sold out their rooms (often at far higher than the norm prices), and *ad hoc* Airbnbs sprang up all over the city (sometimes for as much as $750 a night).[373]

Every store in town, meanwhile, was selling—and often selling out of—the special eyeglasses all the eye doctors told people to have on hand before they dared look up at the sun that day. Among other specially designed look-out spots, SIU campus also planned to open its giant stadium, for those who wanted to have a big, totally unobstructed view of the occurrence.[374]

Leading up to the big day, a local concert promoter hired rock legend Ozzy Osbourne to perform a concert and sing his song "Bark at the Moon" at the exact moment of the eclipse. Area stores sold all sorts of eclipse merchandise including T-shirts, posters, postcards, stickers, and keychains. A company produced an "eclipse soda" (said to taste a bit like Dr. Pepper) and a local winery brewed a special "eclipse beer." Local bakery Cristuado's sold eclipse cookies. NASA and every major TV network and cable news outlet were going to show up, too.[375]

Bucky would have loved the synergy of the sun and the moon intersecting right over Carbondale, and those now in charge of his iconic local home wanted to make the most of this rare occurrence. So, many months in advance, the board began meeting and discussing how they could, too, make the most of all this moon madness. It was decided that on August 19 and 20, as people poured into Carbondale from all over the world, the Dome would be open for tours from 10 a.m. to 4 p.m.

Thad Heckman worked with the city to realign the bus routes to get people to and from the Dome; Lindy Loyd was put in charge of tracking all visitors to the Dome on those two days; Kathy Livingston and Jon Davey worked in the development of possible merchandise; and Larry Weatherford went to work to make minor repairs to the outside of the Dome, especially its primary door (which needed priming and finishing) before the eclipse crowds arrived. Ben Lowder worked on getting the website ready, and Judy Ashby compiled the global "to-do" list that governed everybody's efforts.[376]

The day of the eclipse, Ashby was the one stationed downtown alongside the well-traveled mini-dome where she encouraged all passerby to walk a few blocks over to tour the Dome. Ashby also crafted a series of big signs to plant along the way guiding people to Bucky's house. There, Jon Davey acted as docent and was later spelled by Ben Lowder. Though the tours of the Dome were free, donations were encouraged.[377]

The final take that day from donations and sales of a special, commemorative T-shirt, designed by John Davey's son, Jarrod, showing Bucky's face in profile in front of a rendering of an eclipse, amounted to $2,300. It was enough to purchase a new gate to close off the driveway.[378] That year, all public events surrounding the Dome were tabled including the annual Dome Days celebration in order to focus on eclipse activities.

Since the completion of the outside of the Dome (minus, of course, those ever-troublesome skylights), board members were getting more and more anxious to move inside and start the work necessary to be bring the indoors of Bucky's house back to its original state. Bit by bit, thoughts and actions were moving inwards. Back in April 2016, Jon Davey ran into Evelyn and Cindy Calcaterra, previous visitors to the Dome and, in passing, it was mentioned that the NFP was searching for a vintage refrigerator.[379]

Ironically, the ladies had a working model of a 1950s/early '60s GE refrigerator. Although not an exact match, this vintage "ice box" would serve as a good substitute until a duplicate of the original Hotpoint model could be found. The replacement of fridge was something of a start for all the work that was needed now to make the inside of the Dome look as good—and as original—as the outside now looked.

The new "old" refrigerator.

Yet while the new fridge was great, to really progress with the house, a new influx of funds was desperately needed. The board had been fundraising for over twenty years now; how much more could they continue to appeal to the already very generous local population? Additionally, times had changed. In the late 2010s, Southern Illinois University at Carbondale saw its student population plummet. Before the end of the decade, it was estimated that SIUE, once considered SIUC's little brother or sister, would have a larger enrollment than its Carbondale counterpart.

Problems with state politics (and politicians), with university personnel changeover, and with shifts in U.S. demographics resulted in a steep decline in students that, then, resulted in a major decline in the economics of the city. Ripple effects from this downturn were felt throughout Southern Illinois and would eventually affect the preservation efforts of the Dome.

However, if, over the years, the Bucky Dome and the "Bucky Board," as it has often been termed, had encountered its share of problems and misfortunes—from collapsing roofs to El Derecho to those infuriating skylights—in 2017, something really good happened. Out of the "Bucky blue," an anonymous donor contacted the board with the offer of a very considerable, one-time financial donation—a donation in the upper five figures.[380] If the "Save America's Treasure" grant was the group's—and the Dome's—first great "trim tab" moment, then this was surely its second.

Suddenly, the resurrection of the Bucky Dome, relatively stagnant for the past twelve months, found its second wind, and with these funds, the Bucky Dome would soon be rolling again.

Buoyed by their good fortune, board members began to make their next, imminent plans while the NFP chief "money man," Stephen Schauwecker, worked with the donor's representatives to iron out the details of acquiring the funds.[381] Of course, though certainly appreciated and welcomed warmly, the money brought with it its own set of issues.

Though the original sums were given with no strings attached, the donor did suggest that the money (at least, initially) might be used as leverage in obtaining a matching grant. It was also suggested, by one or two board members, that the money be used simply ("simply") as collateral against which to borrow money via a bank loan—but, from that idea, there quickly arose another question: how would the NFP and the Dome ever generate enough money to pay that loan back?[382]

Many of these topics were still being debated when the funds were given to assist in the preservation and restoration of Dome. Hence, it was decided, with the money in hand, to go ahead and start carefully spending some of this new endowment.[383]

In order to get the most out of their donation, the board immediately voted into existence a subcommittee to oversee the handling and spending of the actual dollars. This subcommittee consisted of Bill Perk, Larry Weatherford, and Steve Schauwecker. One of this subgroup's first actions was the establishing of a separate bank account for these funds, one with its own dedicated debit card.[384]

Meanwhile, at this same time, a second subcommittee was formed just to oversee the about-to-start inside construction work. This subgroup consisted of Schauwecker, Weatherford, Ed Cook, and Thad Heckman.[385]

Since the monies came from a private individual and not from a government grant, the dollars could be spent rather quickly and easily; the board could hire who they wanted without each job having to go out to bid.

As one of the world's best-versed dome experts was already on the board, hiring the lead guy to do the job was easy. Ed Cook would get the nod, pending board approval. Along with Cook's proximity to the project—he lives locally, of course—and his long history with domes, he was also willing to give the board a major "deal" on his services, charging them far under his regular hourly fee.[386] Hence, at a spring 2018 board meeting, all board members present at that meeting, voted "yes" on hiring Ed. There were enough votes for a quorum and the resolution passed.[387]

Shortly after the signature and countersignatures of the construction contract between Ed Cook and a rep from the NFP, a "Notice to Proceed" was issued by the board on June 25, 2018 and it noted an "official" work-start date of the next day.[388]

Joining Ed Cook in the work was Eric Grosshenrich. Later, John Kelly Shannon joined the team and Ed's wife, Nancy Russell-Trude, also came aboard. Grosshenrich, a second-generation builder, carpenter, contractor and jack-of-all-trades, practically grew up on construction sites. He says, "There's photos of me at age seven holding up panels for my dad at worksites." Eric's dad knew and often worked with Ed, and Eric has followed suit.[389]

Eric's first introduction to Bucky Fuller came, be believes, when he was a boy and went to the Science Center in St. Louis. There, they had an exhibit devoted to the famous dome builder. His first introduction to working on domes came by way of his work on some of Carbondale's other domes, including two conjoined domes that sit over on the city's Springer Ridge Road.[390]

Despite the Bucky Dome's high profile and odd, inside angles, Grosshenrich has largely found the work a "typical remodeling job," a beautiful home that has "not been taken care of."[391] Like Cook—and quite generously—the other two men worked for an hourly rate far below their norm in order to show that they, too, were committed to saving the Dome.

On the morning of June 26, 2018, Cook and Grosshenrich pulled their trucks into the driveway of 407 South Forest, to begin work. Along with his tools and considerable expertise, Eric Grosshenrich also brought something else with him—a small digital camcorder easily affixed to the brim of his ball cap. On his first day, Grosshenrich began to make daily short videos charting the Dome's progress and then posting them on YouTube.

Narrated by Grosshenrich in a funny deadpan style (he says in a couple of them when he happens on something odd, "Well, that's not supposed to be there"), the videos, when viewed in sequence, provide the curious with an "insider" look at the house and the surprises and pitfalls of home restoration. They begin as follows:

Day 1: Intro and tour of the Dome's interior.

Day 2: An introduction to compatriot Ed Cook and a look at the home's existing wiring.

Day 3: A second set of drywall? Looking behind the refrigerator, Eric discovers that the drywall there is not the home's original stuff.

Day 4: The insulation arrives and the ceiling of the bedroom (also not original) begins to be removed.

Day 5: Duct work!

Day 6: Installation of insulation starts, necessitating the removal of additional internal panels; a rogue pine cone is discovered beneath one panel, it has been sealed up in the wall for more than four decades.

Day 7: The scaffolding arrives so that the inside top of the Dome can be reached by Ed and Eric;

Day 8: Electrical work begins.[392]

When the men first arrived at the Dome, they arrived with a daunting "to-do" list:

Preservation or replacement of the home's original, interior, triangular wall panels, either those that were severely rotted away or those that were poorly-configured repairs;

Preservation of the hot-water circulating boiler and related although Mike Mitchell had replaced the original boiler about a year after he moved in in the 1970s;

The installation of vent fans in each of the home's two bathrooms, per current Illinois building codes.

And air conditioning. Originally, the home had no AC.

The house's basic design along with its central ceiling fan and the lower temps of decades ago helped keep the home cool. Additionally, originally, the home was only really meant for two people—Bucky and Anne. But, with the possibility of large groups (perhaps as many as 25 to 30 at a time) now coming to the Dome, it was decided that AC should, for everyone's comfort, be installed.[393]

The Dome as construction site; Ed Cook is seen on the scaffolding.

So, together, Thad Heckman, Ed Cook, and Ed Grosshenrich investigated various AC options while also worrying about some very basic questions about how one retroactively installs AC into a home that never had it before. Where does the outdoor unit sit? What about ducts? Can the necessary indoor machinery fit in the Dome's relatively small utility closet located in the center of the home?[394]

Ultimately purchased was a Pioneer brand RYB-22 unit. The necessary outside AC condenser was nestled, as inconspicuously as possible, into a space between the back of the house and the driveway; when the fence is eventually restored, the fence will conceal it.[395]

Inside, the necessary equipment was positioned in the ceiling above the home's existing utility closet. To support the weight of the unit, brand-new steel reinforcements were installed. Grosshenrich measured the space in the utility closet carefully and then welded together the steel bars to make the perfect support and framing device. He welded the pieces together in the driveway, his welding torch powered by solar energy harnessed from reflectors on the roof of his trailer.[396] The final cost for the AC unit and its installation was a cool $7,500.[397]

Though taking the home back to its 1960s state was the NFP's primary purpose, central air conditioning was not the only "modernization" that was made. Some of the changes were based upon newer safety requirements—for one, smoke detectors. Also, the original glass in the home's sliding doors was plate glass, but plate glass is considered too dangerous today due to its shattering and lacerating tendencies. It is a major "walk-through" hazard, and more than once in recent years, visitors to the Dome have accidently walked right into one of the doors. Thankfully, it did not shatter, but what if it had? So, plate has been replaced by tempered glass. As Thad Heckman has noted, "Sometimes safety outweighs originality." The new glass not only is far safer but also more energy efficient, performing far better than what Bucky could have imagined in his day.[398] Today, the only glass in the home that is 100 percent original is the glass in the two windows by the front door on the home's east-facing side.

Additionally, the home needed to be fully accessible. Though the Dome's front door is too small for a wheelchair, any of the home's sliding doors would permit entry. Meanwhile, the construction of accessibility ramps are also among the Dome's plans.[399]

As relatively straightforward as these various tasks seemed to be in the planning stages, in true Bucky Dome fashion, once everyone began to look more deeply at the house, new issues suddenly presented themselves, as did a plethora of head-scratching questions. Very soon, it was discovered that for a home that is relatively young (especially compared to the other homes in its neighborhood), many alterations had been made to it over the years.[400]

A case in point—behind one of the panels in the bedroom, a large metal conduit (a pipe of sorts) was found running up the wall and containing the wires from an electrical system upgrade done some years after the Fullers moved in. Or at least that was what Ed, Eric and Thad have theorized because when the conduit was installed whoever installed it cut away some of the Dome's structural supports, and it is unlikely Bucky would ever have allowed that to happen.

Yet what of other changes? Were these made by Bucky? Fuller loved to tinker with his inventions. Might he had treated his own home—perhaps at times to the consternation of his wife—as just a giant laboratory? Perhaps, but perhaps more realistically, it was during the two-plus decades that Mike Mitchell owned and rented out the home, but

The rogue conduit discovered behind the bedroom's wall.

resided on the West Coast, that augmentations were made to the home, either with his consent or even without.

Regardless, for the group trying to bring Bucky's house back to its original state, it meant nearly daily surprises and investigations, what the long-time board member Judy Ashby has called "modern-day archeology."[401]

Some of the revelations discovered were not so great. One of these was the first evidence of termite damage to the home; it was found when the interior wall that divided the living room from the master bathroom was peeled away. Along with those vermin, the space between the walls were, as Ed Cook put it, "not good if you have arachnophobia."[402]

A few oddities appeared as well. Sometime during the home's lifespan, its central thermostat was moved. For years, the thermostat was just inside the front door but once a wall panel in the living room was removed, evidence of the thermostat originally being in the living room was unearthed.

Other wiring curiosities were also discovered. Electrical switches at the base of the home's central staircase were always thought to be for the home's central, suspended fan, but when investigated, it was discovered that instead, these switches were actually linked to the two sets of electrical outlets up in the home's loft floor.

However, the "fishing" for wires through the walls was just the tip of the iceberg when it came to the electrical aspects of the home. Exhuming the walls uncovered a disturbing jumble of scorched wires, flaked-off wire casings, and various burnt or singed areas. There were even examples of nails that had been pounded into the wall and pierced through existing wiring. Heckman has said, "It's a miracle this thing hasn't burned to the ground."[403]

Eventually, every inch of original wiring—some containing the "worst electrical insulating I've ever seen," according to Eric Grosshenrich—would have to be replaced.[404] The new, bright yellow wiring, once put in place, was, in the words of one board member, "beautiful."[405]

Almost as surprising as the compromised, dangerous wiring, they found in the house was what they found outside the house. One day, Ed and Eric went looking for how to turn on the outside fountain. They had to dig up some of the yard for that one, but they did eventually find an underground cable that extended from the home out into the yard. However, surprisingly, the wire did not go to the fountain but to a previously unknown of light, embedded into the yard, just inside the fence, located at the home's southwest corner. No doubt, the radiating light was designed to create lovely shadows in its interplay with the water of the fountain. The fountain, it was discovered later, could only be turned on by the Fullers at the actual fountain pit.

For information about the previously unknown light, the team once again reached out to Mickey Mitchell, the Dome's second owner, for details and, if possible, a photo or two. As always, Mickey provided ample insight.

Along with wiring and conduits in that direction of the house—southwest—the wires that powered the house's two outdoor, driveway-adjacent bollard lights had to be replaced and were strategically placed to avoid cutting up the driveway.

Meanwhile, alarmingly, an exploration and removal of the ceiling in the kitchen and bedroom revealed that there was very little actual structural support for the home's loft area. Though some wood crossmembers were in place, the extreme ends of each side of the loft had very little attachment as far as support, causing the loft to sag in some places by as much as a full inch. When discovered, a somewhat belated panic ensued as many recalled all the big tour groups, boxes of books and other heavy items that had at times been dragged up there.

Six steel reinforcement brackets were eventually invented, fabricated by Eric and welded together. These rectangular steel frames were installed as under-the-floor supports. It was a red-letter day when, after this work was done, Ed and Eric took out their level to see if the floor of the loft was actually squared and flat. The shot of the bubble dead center in the level called for a dramatic zoom-in and close-up from Eric's camera. The moment was preserved for posterity in a video they posted on YouTube.[406]

While all these discoveries were good to learn, and even interesting in a way, they came, quite literally, at a cost. Anything unexpected meant that new work had to be done, more time would be needed for preservation, and new or other materials had to be purchased. These items had not been budgeted. And though the board had, now, obtained a considerable nest egg, they were soon burning through the gift far faster than they had planned. The unexpected and the overruns (common in major preservations) were so frequent and, in the end, costly, that by the end of 2018, even the board's construction contingency fund had been nearly exhausted.[407]

View of the ditches dug pursuing the underground wires found in the yard.

Screwed into place: one of the newly fashioned metal reinforcements created to support the upstairs loft.

Still, despite the higher than expected burn rate, all the members of the board hoped that their coffers would hold out so that Phase II of the restoration, still planned to end in early 2019, could be finished. The board would assist this by doubling down on their fundraising efforts which included at this time a city of Carbondale grant application applied for during the fall of 2018.[408]

Back inside the house, the installing of the home's new insulation was an especially interesting undertaking. The home's original 1960 insulation were sheets of the typical fluffy, cotton candy-looking stuff (though yellow colored) that most homeowners are familiar with today. On the day of the build in 1960, the insulation arrived already adhered to the backs of the triangle panels, adding to the efficiency of the home's construction. On top of the insulation, facing in towards the interior of the home, there was also a thin sheet of foil also typical of home insulation at the time. Though the original insulation was about 1 to 2 inches thick (better than normal for buildings at that time but low by today's requirements), the NFP decided that a thicker insulation should be installed as part of this restoration.

A suggestion of spray-on insulation was favored but rejected as, if it ever had to be removed, pulling it down could damage the original Dome pieces. Approved by local authorities, the new insulation assembly finally chosen was a 2.5-inch thick solid closed-cell foam insulation with fiberglass facing insulation that was purchased in large rectangle sheets measuring 4 feet by 8 feet. For the conventional home, installing these insulation sheets would be a breeze but, for a geodesic dome made up of a series of interlocking triangles, each panel had to be cut by Ed Cook and then fitted in place. Luckily, all the triangles were about the same size, so once Cook determined the proper lengths, it was smooth sailing.

After each triangle was cut, they were wedged into each interior slot between the home's wooden frames. Then each was further held into place with expanding sealant. Emerging from the worker's caulk gun as an orange-y goo that looked a bit like Cheez Whiz straight from its can, this caulk would then expand as it dried, capably filling in, solidly, every crevice between insulation sheet and the wood. In fact, the caulk expands so much it often bubbles out. Hence, after it is completely dry, workers then have to return with knives to cut off the excess, thus leaving small flakes and snakes of orange foam to be swept up later.

As Thad Heckman has stated, "Internal moisture—any moisture—is the enemy of a home, any home." Hence, even after the application of the caulk, and after the caulk "cured," Cook went back to each seam and filled in more caulk to further seal any tiny pocket or niche where water could gather. That effort, along with a new application of seaming tape over the slender cracks between each board, would retard any moisture migration. For non-triangle, more vertical areas, or areas not readily accessible, a stone wool insulation was fitted into the walls. The final cost for the insulation was just under $7,000.[409]

Once the insulation was successfully installed, it was time to attach the vapor retarder and barrier. These big sheets of foil provide a moisture barrier and serve as a heat reflector, too. The foil sheets would help retain heat in the winter. Since the barrier arrived in a big roll, it too had to be carefully cut into triangles. To do so, Ed Cook first had to set up a temporary vapor barrier dispensing table, a sheet of plywood set up on some scaffolding. Once each triangle was cut, it was then stapled into place. Yet, as

A wall of the Dome after the installing of the insulation.

noted by Grosshenrich, "A vapor barrier is not a vapor barrier if it has holes in it."[410] So, once put up, all seams and any punctures had to then be sealed over with a special roofing tape. Super sticky and moisture resistant, the tape also seals the staples and the insulation. Grosshenrich's camera captured the installing of the first triangular sheet and him then noting, "Only about a thousand more to go."[411]

By the time the vapor barrier was fully installed, the entire inside of the Dome looked like it had been wrapped in tin foil. After that, a special foil tape was applied to all the seams followed by a second layer of black sealing tape. Together, they will form as close to a 100 percent moisture barrier as possible. In late 2019, a bore hole (necessitated by a new vent) was made and it indicated perfectly clear, clean and dry materials spanning all the layers of the Dome's shell.

The vapor barrier began to be installed on August 27, 2018, and while Cook continued with that task, that same day, Grosshenrich headed outside to cut some of the concrete sidewalk. The original walk—already cracked completely through—had to be removed to install the lines for the new heat pump and to install the new electrical lines for the two driveway-located bollard lights then undergoing restoration by Larry Weatherford.

Day twenty-five of interior work saw the incorporation of new fuses into the old electrical box. The original box has been retained for the sake of historical accuracy.

The Dome with vapor barrier in place.

The original electric box was found only after a second electrical box was removed. The retrofitted box was almost certainly installed during an upgrade in service and was simply placed directly in front of the old one.

Day thirty-five saw the further removal of the ceiling in the kitchen and—as suspected—the big reveal of its rather weak and odd framing, which, though probably enough for a ceiling, was not necessarily strong enough to actually act as a floor to the upstairs study. Modifying and strengthening that floor first demanded the temporary installation of various support beams. For this, the men used "LVLs" (laminated veneer lumber). LVL is specially constructed, reinforced lumber which, as Eric said in his video, makes the wood "stronger than how trees grow it."[412]

That day also saw the start of the replacement of the home's original wall panels. When they were originally removed at the start of the restoration several years ago, each triangle was labeled with a number and then a grid, or key, was made to track where each one had come from so that, eventually, they could be returned to the same place. Unfortunately—as it was probably very easy to do—someone either mislabeled a piece or two (or more) or transcribed the numbers incorrectly. Hence, when the pieces began to be laid out to be put back up, almost nothing fit and almost nothing matched.

So, Ed and Eric, with Thad Heckman's assistance, had no choice but to start from scratch and, like a giant, 3D jigsaw puzzle, attempt to locate where every single interior triangle belonged. To get started, the men first located the pieces cut out for the overhead skylights—each of these could only fit in one place and in one orientation. Using those, then, as a starting point, the guys could then piece together where the remaining triangles had to go.

Before their reinstallation, each panel was first inspected and sanded down. Some (many in fact) of the removed panels were found not to be original to the home at all, but were latter-day replacements installed either by the home's second owner or by one of his tenants. Additionally, some original panels were, sadly, too warped or damaged to be utilized. For their old locations, new triangles had to be created from plywood, something that Ed Cook did on site. As he began his work, Cook quickly discovered the cutting of the panels (both old and new) would not be a quick, one-size-fits-all undertaking. Though the Dome is now in its proper shape—or sphericity—any tiny fraction of an inch that it is off can create a dangerous ripple effect. Hence, each piece is not exactly equal (but close) and many of the new interior pieces had to be cut or trimmed to fit exactly.[413] As Thad Heckman commented once, "When domes fail, they fail spectacularly."[414]

To reattach the panels, both old and new, small finishing nails were used. Small amounts of putty were then set into all the nail divots for a smooth surface. A new coat of primer and paint (white to match the original) was then applied. In the original home, long strips of wooden half-round were used to cover the edge seams between each triangle of the Dome. This trim would be restored as well. Interestingly, Bucky did not use the trim on all of the interior seams so as not to, perhaps, create a busy, "cage-like" feel. That meant that the butt joints between the various interior triangles had to fit together as tightly and neatly as possible.

In April 2018, one of the internet's ever-so-popular lists came out. This one, originating from Kendall County, Illinois, asked readers to nominate the top structures still standing in Illinois. While Chicago's Wrigley Field earned the top spot, Carbondale's Bucky Dome also made the cut.[415]

In July 2018, after considerable research, the board purchased a series of outdoor security cameras to keep the Dome (and all their hard work on it) safe.[416]

In August 2018, long-serving Board President John Davey was one of the recipients of the Lindell W. Sturgis Memorial Public Service Award for his efforts both in the yearly Kids Architecture camps he has run for close to three decades and for his lengthy commitment to the Dome.[417]

In late 2018, up in Edwardsville, Illinois's other Fuller dome was the site of a "Bucky Weekend," which celebrated the donation of one of Fuller's original art print portfolios to SIUE. It was donated by his estate. To commemorate its donation, Bucky's daughter, Allegra (accompanied by her daughter, Alexandra), traveled to Edwardsville at the invitation of Ben Lowder, SIUE staffer and Bucky board member. Mrs. Snyder was also scheduled to travel down to Carbondale to see the progress on her mother and father's old home. Unfortunately, an illness at the last minute prevented her from visiting.[418]

At the start of 2019, everyone was excited about the great progress that the Dome had made in the past few months. Yet, funds were flowing out fast, and though the Dome was starting to look liked its old, original self again, there was still a lot of work to be done. To help them along, new energy came to the board thanks to the addition of three new board members: Anita Presley, Dave Presley, and Christian Baumann. These three

The Dome with original, interior wall panels installed.

Interior wall panels, old and new.

Area of Dome interior which necessitated all new interior panels and close-fitting seams.

Example of some of the original upper, interior half-round trim over some seams inside the Dome.

new members helped fill the vacancies left by Jessica Allee and Peter Szczesniweski, who had both exited to pursue other interests.

Husband and wife Dave and Anita came to the board by way of Daniel's brother, longtime NFP board member Daniel Presley. Having recently relocated to Carbondale from Arizona, the Presley's were anxious to get involved after hearing about the Dome from Daniel for many years and seeing the Bucky-related exhibit at Carbondale's Science Center. For Anita, having just retired from a career in the Air Force, she believed that her particular skillset could be of use in various aspects of Dome business.[419]

Similarly, husband Dave, who had grown up in Carbondale, was interested in getting re-involved with the community. Dave and his wife run two new internet-based businesses, and his knowledge in IT as well as his knowledge of Carbondale history would be of use to the group.[420]

Also joining up at that time was Christian Baumann, an SIUC student, originally from Charleston, Illinois. After beginning his studies at Eastern Illinois U., Baumann later transferred to SIUC for their architectural program. He was in his junior year when one of his professors, Jon Davey, knowing of Baumann's great interest in architectural history, suggested he come aboard.[421]

So far, Baumann's involvement has included everything from carving one of the pumpkins for the annual display (he did the Lighthouse of Alexandria) to proofreading a gallon of text from the group (Christian says, "I have sort of a knack for that; I'm a little OCD") and also serving as an important bridge between the university's student population and the NFP.[422]

The new members arrived at a time that was quite exciting. As is often the process with construction, intense early progress with demolition and other plainly visible tasks is followed by a long stretch of necessary but hard-to-see progress involving electrical, plumbing, and other "behind-the-scenes" items. This, though, is then followed by a period of seemingly rapid progress with walls and finishings being reinstalled, making all the hard work very "see-able." Hence, by February 2019, when the internal wall panels were up and primed and painted, the Dome was looking like the Dome.

With the walls restored in the living, dining and bedroom areas, it was time now for the board to take on the bathrooms. Luckily, the basic plumbing of both baths was in pretty good shape; just some spit-polishing is needed on the fixtures. Not so good however were the vanity and the shower stall in the smaller, "his" bath. At some point, someone had completely recovered the original tile inside the shower. The covering—very long sheets of vinyl with adhesive on the back—was pulled down easily but left behind a series of ugly, "S"-shaped glue marks all over the porcelain. Though it did take a lot of elbow grease to scrub them away, they could, thankfully, be fully removed.[423]

Meanwhile, that shower's plastic shower door is probably original but is also cracked and yellowed; it will need to be replaced. The medicine cabinets in both baths are believed to be original but the three-bulb vanity lighting mounted above the mirrors in both turned out to be too recent in vintage. The board then went about doing some research about what the lighting might have looked like in a mid-century modern home. Then fortune smiled upon Thad Heckman one day when he was in nearby Murphysboro and found a trove of retro-looking features which the owner said was there for the taking. One fixture—all glass, curved, and frosted—looked liked how a bathroom light might have looked in the late 1950s–early 1960s.[424]

The newly re-installed panels, with trim, and freshly painted.

As the bathroom work commenced, the team also devoted its energies to some of the restoration's finer details. After pouring over all the old photos from the days when Bucky and Anne lived in the home, Thad Heckman came up with a list of items that needed to be recovered, located, and reproduced. He assigned all the board members to a specific task and encouraged them to visit other job sites, estate sales, antique stores, and anyplace else they could go to find some of the original items or something close to it.[425]

Larry Weatherford was to refurbish the driveway lights and Daniel Presley was tasked with finding an appropriate storm door. Thankfully, the Fullers' original aluminum door was "nothing special," probably purchased during the original build from a local hardware store. Judy Ashby took on the light switch covers while Jon Davey concentrated on the bathroom fixtures. Bill Perk and Lindy Loyd had to find that three-bar towel rack that Anne had installed above the kitchen sink. Ed Cook was in charge of finding or replicating the home's large central fan. Heckman had already located a pencil sharpener like the one Bucky had near the bedroom door. Everyone was to keep their eyes peeled for a match to the original refrigerator and a match to the telephone like the one originally in the bedroom.[426]

Then, in early 2019, the board was contacted by Bucky's daughter, Allegra Fuller Snyder. Despite not being able to see the restoration in progress, Mrs. Snyder did say that she was in the process of gathering up all of her parents' old possessions, originally in the Dome which she had inherited, for donation back to it. These items included the iconic Fu Dog statues, the Eames chairs, Anne's writing desk, and numerous books from Bucky's personal library. The board welcomed this generosity and board member Ben Lowder, who has his own large panel truck for the transporting of his large sculptures, volunteered it (and him) for the eventual cross-country road trip to retrieve the artifacts.[427]

The Fullers' original storm door, *c.* 1966. (*Courtesy: Tony Gwilliam*)

The Fullers' original Fu Dogs.

Sadly, one iconic item that will not make it back to the Dome's interior—at least for now—is the Ruth Asawa amoeba-like hanging wire sculpture. Though still in the possession of Allegra, it is in too constant demand all over the world for Asawa retrospectives.[428]

One unexpected addition to the home will be some of Bucky's own clothing. Herb Roan's daughter, Toni, years ago inherited a handful of Fuller's personal items and has kindly gifted them to the Dome. They include two monogramed shirts, two bow ties, and a pair of suspenders. They should eventually look quite charming inside the bedroom closet.[429]

Meanwhile, the upstairs loft railings were reinforced, and research began on sprucing up the kitchen cabinets and on replacing the cork on the home's main floor.[430] In regard to the latter, the cork company WE Cork was consulted again about the best way to revive and treat the existing cork flooring in the home. According to company owner Ann Wicander, back in the 1950s and early 1960s, cork floors like those that were put down in the Bucky Dome used to be treated with beeswax. Today, polyurethane sealers are suggested. These sealers are applied after the cork is lightly sanded to remove any stains or rough patches. The poly then, obviously, seals the cork and the floor can then be buffed to a high shine.[431]

As all of this work continued, it was also time to revisit an old foe: the skylights. Even after various earlier "repairs," those overhead windows remained worrisome—not egregiously but enough to raise concern. The new approach, arrived at by the team, was multi-layer, properly insulated glass set in a full bed of sealants, rated for direct skyward exposure and ultraviolet protection then treated with an additional layer of sealant over the joints between the roof and the skylight itself. However, as Ed Cook and Thad Heckman noted, even this fix was probably not foolproof. Today, few builders would install skylights flush with the roof. While none of the lights are leaking as of this writing, permanently warding off Mother Nature is a nearly impossible task, and like periodically repainting the walls, skylight up-keep will just be a part of Dome's ongoing maintenance.[432]

Of course, all this work—as exciting as it was—did not come without a price tag attached. So, the NFP kept on fundraising. In April 2019, the Dome was the recipient of two bits of good news from the city. First, it was awarded Carbondale's 2018 Historic Preservation Award for all the progress made.[433] Additionally, the Dome obtained a Mayor's Grant from the city for $5,000.[434] In June, a yard sale of donated knick-knacks held at the Dome brought in (rather surprisingly) an additional $5,000.[435]

This influx of funds coincided with a NFP website redesign that was spearhead by board member David Presley. Along with the site's overall functionality being simplified, its "purchasing" options—where visitors to the site could buy Bucky Dome T-shirts, the *Roam* CD and copies of Thad's Holland drawing, among other items—was beefed up. Then, several members began to work on a photo array and video presentation to form the bedrock of a GoFundMe campaign.

Still, despite all the momentum, progress and donations, more funds were still needed. The cork floor and its installation alone will cost in the neighborhood of $10,000. Both bathrooms still need to be completed. While the kitchen cabinetry is now reconditioned and the newly reinstalled range hood works, the kitchen sink needs to be replumbed and its disposal repaired. Then there is the refurbishing of the home's original accordion

The finally finished skylights.

door and the locating, or fabricating, and installation of its central industrial-size ceiling fan to take place, hopefully, in the near future.

Outdoors, there is the reinstallation of the bollard lights and the fixing of the fountain—the latter could run the board as much as $5,000. Then, of course, there's the notorious Bucky Fence. Only a portion of it is still standing, and though many of its original slats have been saved, many others will have to be completely replaced. Also, redwood is one of the more expensive of woods.

Sadly, as 2019 wore on, despite the best efforts and hopes of many, the Dome's two biggest, most time-honored foes—time and money—scotched the board's original hopes of having the home ready (for at least a "soft" opening) by the end of 2019. Though, as of this writing, it is close.

Still, hopes remained—and remain—high that the necessary monies can be obtained soon and that future construction "surprises" will be minimal, and that the critical path forward will be a short one. For as Bucky himself once sang:

> *Let architects sing of aesthetics that bring*
> *Rich clients in hordes to their knees*
> *Just give me a home in a great circle dome…*

Above: The loft.

Below left: The entryway today.

Below middle: The living room, the Bucky Dome, late 2019.

Below right: The renewed living room.

Above: The kitchen, restored.

Below: The Dome today.

AFTERWORD:
EXO- (2020 ONWARDS)

exo-: ex-o (/'exo/) outside; external.

Carbondale's beloved "Bucky Dome" was put up in one day in 1960; it has endured for over fifty years. In its over half-century of existence, it has been the possession of a handful, the home to about a dozen or so, and the obsession of a few. In its survival to the present day, it has had more than a little help.

Can a building be classified as "ephemera"? Bucky himself might have viewed his one-time home in such a way, but for the sake of history, it is imperative that such a structure, so much the epitome of Fuller's building and life philosophy, be sustained and be made to endure—forever, if possible. Ironically, it is doubtful that the Carbondale Bucky Dome would ever be built today, at least not in its current location. City codes would prevent it for not being in keeping with the "appearance of other homes in the neighborhood."

The preservation of many of Fuller's other most important works have not fared nearly as well as his one-time Illinois home. Only one of his three original Dymaxion Car prototypes still exists. Although his Montreal Biosphere, for 1967's World's Fair, is still, technically, standing, only its actual "skeleton" remains; the sphere's outer "skin" was burned off due to a welding mishap in 1976.

Still, what remains of the Biosphere is more than what remains of 1958's Union Tank Car Dome in Baton Rouge, Louisiana. Originally used as a repair station for railroad tankers, the building was later abandoned by Union when tank cars increased in size and made further use of the building impossible. In 2007, after years of neglect, but amid public outcry for its preservation, that structure was demolished.

The sad demise of many of Fuller's greatest works should not, perhaps, be of enormous surprise. Despite their innovation and appeal, at the time that these other constructions were in need of preservation, Buckminster Fuller had not yet truly come into his own full status as the visionary that he was. One-time Bucky Dome owner Bill Perk proved prophetic in 1985:

It would not surprise me if it took another thirty to fifty years before the full significance of Bucky's ideas became more widely visible, at which time people might think we should have done something to recognize him more.

Cornelius Crane, one-time Bucky Dome board president, once struck a similar chord. He said of the home, "We just believe it's just an absolutely important artifact. Maybe 100 years from now, they'll thank us and say 'Thank goodness they saved that Dome.'"

As of this writing, the restoration efforts of the RBF Dome NFP continue on behalf of the Dome and its rebirth. Phase I has been completed; Phase II, interior preservation, is almost (almost) done. Phase III, "Grounds," is pending. Phase III will bring the outside yard, fountain, driveway and the infamous Bucky fence back to their *c.* 1960 state. Then, already somewhat in progress, is more recently established Phase IV, which is devoted to the full restoration of interior furnishings, done to replicate every inch and artifact of the Fullers' original décor as closely as possible.

Then, there will be an ongoing, final phase that will, among other activities, see to the full utilization of the Dome as a historical and educational site. The gathering of money (which includes all proceeds from this book) is to guarantee the long-term care of the Dome. Off in the distant future is the creation of a Fuller interpretive center to be erected nearby so that visitors to the house can get the most out of their visit to this great Bucky landmark.

Ideally, of course, this book would end with the end of the first four phases. However, as NFP board members and the people of Carbondale can tell you, not everything always goes to plan. As former Bucky board member Janet Donoghue once surmised, "It's a relay race."

The "finishing" of the Bucky Dome is an ongoing endeavor that still requires energy, commitment and fortitude. In some ways, ending this book with the Dome still in development echoes an important Fuller construct: everything is always a dwork in progress.

APPENDIX: EXTRA-

extra-: ex·tra (/ˈekstrə/) adj. in addition.

Dome Ownership History

R. Buckminster and Anne H. Fuller	8/1960–9/1973
Mike S. Mitchell	9/1973–July 1999
Bill Perk	July 1999–2002
RBF Dome NFP	2002–present

"Livingry": A Tenant and Rental History*

R. Buckminster and Anne H. Fuller	8/20/1960–9/1973
Brendan O'Regan	c. 1973
Mike "Mickey" Mitchell	1973–c.1976
Michelle Bach	1976
Linda Burzynski	1978–1979
H. B. Koplowitz	9/1979–11/1983
Gary Marx	1979–1980
Deborah E. Browne	"late" 1981–11/1983
Donald S. Wham	1983–1985
Paul Hensel	1985–1986
William Clay	1986–c. 1987
Linda Buchholz	1988–1995
Pete DePaoli/Kim Alexander	June 1995–Dec. 1996
Colleen M. Doyle	1997
Martin Moran/Matt Staulcup	August 1997–1998
William Denton	c. 1998
Alexis E. Holle/Troy C. Smith	1998–1999

Kevin J. Connor   1999–2000
Justin R. Edgren   *c.* May 2001–Summer 2003

*Rental records—and people's memories—from these years are somewhat patchy; this list of one-time Dome tenants is as complete and accurate as we know at time of writing.

RBF Dome NFP Presidents

Cornelius Crane   7/16/2002–2009
Brent Ritzel   2009–1/24/2012
Thad Heckman   1/25/2012–2/13/2012 (interim)
Jon Davey   2/14/2012–present

ENDNOTES

Chapter 1

1. The early history of Southern Illinois University at Carbondale (SIUC) can be found at: siu.edu/about-siu/history.php

2. University Archives, Morris Library, Southern Illinois University at Carbondale: lib.siu.edu/collections/scrc/scrc-collections/university-archives.php

3. Blades, J., "Russo's no fool," *Chicago Tribune* (Feb. 19, 1995), p. 171.

4. Hatch, A., *Buckminster Fuller at Home in the Universe* (New York: Dell, 1974), p. 218.

5. A brief biography of Fuller can be found on the homepage of the Buckminster Fuller Institute's website: BFI.org (bfi.org/about-fuller/biography)

6. Baldwin, J., *Buckyworks* (New York: John Wiley, 1996), p. 40.

7. Schnyder, M., "National history museum showcases little-known Wichita artifact." *Wichita Eagle* (Dec. 19, 2016).

8. Potter, R., *Buckminster Fuller* (Englewood Cliffs, NJ: Silver Burdett, 1990), p. 104.

9. *Ibid.*

10. Baldwin, *op. cit.*, pp. 112-113.

11. *Ibid.*

12. Olive, R., "Dome Provides Light Structures," *Eugene Guard* (Eugene, OR) (Jan. 18, 1959), p. 26.

13. The website of the Buckminster Fuller Institute, BFI.org, has a chronological list of all Fuller-obtained U.S. patents: bfi.org/about-fuller/bibliography/patents

14. Baldwin, *op. cit.*, p. 110.

15. "'Bucky' Fuller, Design Revolutionary, Still Super-Charged with Ideas," *Southern Illinoisan* (October 26, 1959), p. 16.

16. Buckminster Fuller Institute (BFI.org): bfi.org/about-fuller/biography/fullers-influence

17. Baldwin, *op. cit.*, p. 166.

18. "Cambridge City Area First with Geodesic Dome Dealership," *National Road Traveler* (Cambridge City, IN) (Sept. 24, 1959), p. 1.

19. Pease Domes brochure, Pease Manufacturing Company (Hamilton, OH), 1959.

20. *Ibid.*

21. *Ibid.*

22. Hatch, *op. cit.*, p. 218.

23. Powell, L., "An engineering whodunit at Black Mountain College," *North Carolina Miscellany* (Oct. 13, 2017).

24. Hatch, *op. cit.*, p. 218.

25. *Ibid.*

26. *Ibid.*

27. Interview with Harold Cohen, Nov. 20, 2015.

28. "SIU Appoints Fuller," *Southern Illinoisan* (Sept. 20, 1959), p. 3.

29. *Ibid.*

30. Mitchell, B., *Delyte Morris of SIU* (Carbondale, IL: SIU Press, 1988), p. 94.

31. "The Dymaxion American," *Time* (Jan. 10, 1964), p. 46+.

32. *Ibid.*

33. Gelman, B., "Fuller sees great progress toward 'world man,'" *Southern Illinoisan* (Sept. 24, 1967), p. 36.

34. "The Dymaxion American," *op. cit.*, p. 46+.

35. Gelman, B., "Tired? Just think about Bucky's schedule," *Southern Illinoisan* (Apr. 27, 1967), p. 3.

36. Rush, L., "Always on call," *Southern Illinoisan* (April 22, 2010).

37. Interview with Shirley Sharkey, April 17, 2016.

38. *Ibid.*

39. Applewhite, E. J., *Cosmic Fishing: An Account of Writing Synergetics with Buckminster Fuller* (New York: Macmillan, 1977), p. 12.

40. "The Dymaxion American," *op. cit.*, p. 46+.

41. Hatch, *op. cit.*, p. 247.

42. A chronological list of all books published by Fuller can be found at the Fuller Institute's website: BFI.org bfi.org/about-fuller/bibliography/books-by-fuller

43. "Scenic Design Contest Planned," *Southern Illinoisan* (Jan. 7, 1962), p. 19.

44. "Fuller Will Talk Jan. 11," *Southern Illinoisan* (Jan. 3, 1963), p. 2.

45. "Fuller to Talk on World Design," *Southern Illinoisan* (May 12, 1964), p. 3.

46. "Art Educators Meet at SIU," *Southern Illinoisan* (Nov. 1, 1964), p. 3.

47. "Fuller to Talk Monday," *Southern Illinoisan* (Feb. 7, 1965), p. 2.

48. "SIU Business Writers Meet Starts Today," *Southern Illinoisan* (Apr. 9, 1965), p. 2.

49. Gelman, B., "Let's Play 'Make the World Work' Fuller Urges," *Southern Illinoisan* (Mar. 10, 1965), p. 3

50. "Buckminster Fuller Will Speak to NUEA," *Edwardsville Intelligencer* (Oct. 25, 1967), p. 9.

51. "Fuller slated PTA speaker," *Southern Illinoisan* (Sept. 7, 1967), p. 7.

52. "'Bucky' to picnic with area children," *Southern Illinoisan* (May 26, 1971), p. 2.

53. Cohen interview.

54. "Dome Home For Fuller," *Southern Illinoisan* (Oct. 21, 1959), p. 1.

55. E-mail communication to the authors from Maureen Berkowitz, Jackson County Assessment Office, June 23, 2015.

56. O'Dell, C., *Virginia Marmaduke: A Journey in Print from Carbondale to Chicago* (Chicago: Arcadia, 2001), p. 13.

57. Interview with Ira Parrish, Oct. 1, 2015.

58. White and Borgognoni Architects, "Buckminster Fuller Dome Home Evaluation Report" (Carbondale, IL), 2005.

59. *Ibid.*

60. Interview with Thad Heckman, July 5, 2015.

61. *Ibid.*

62. *Ibid.*

63. *Ibid.*

64. *Ibid.*

65. *Ibid.*

66. Pease Domes brochure, 1959.

67. "'Haystack House' Termed Coming Thing in House," *Hartford Courant* (May 15, 1960), p. 10D.

68. Pease Domes brochure, 1959.

69. "Dome Home is Erected in Seven Hours," *Southern Illinoisan* (Apr. 20, 1960), p. 3.

70. Parrish interview.

71. *Ibid.*

72. *Ibid.*

73. Heckman interview.

74. Parrish interview.

75. *Ibid.*

76. *Ibid.*

77. *Ibid.*

78. *Ibid.*

79. *Ibid.*

80. From late 2015 until its publication, Mike "Mickey" Mitchell, the second owner of Carbondale's "Bucky Dome," exchanged hundreds of e-mails with the authors. This information is gleamed from that correspondence.

81. Parrish interview.

82. *Ibid.*

83. *Ibid.*

84. *Ibid.*

85. Heckman interview.

86. Parrish interview.

87. Heckman interview.

88. Parrish interview.

89. *Ibid.*

90. "Dome Home is Erected…", p. 3.

91. youtube.com/watch?v=wltL_qdj3-s

92. "Dome Home is Erected…," *op. cit.*, p. 3.

93. Interview with Herb Meyer, Sept. 18, 2015.

94. Heckman interview.

95. "The Dymaxion American," *op. cit.*, p. 46.

96. "'Haystack House' Draws Wide Interest," *Southern Illinoisan* (May 27, 1960), p. 3.

97. "Dome Home Number Two Is Going Up," *Southern Illinoisan* (May 10, 1960), p. 2.

98. "Domes for Designers," *Southern Illinoisan* (Dec. 6, 1960), p. 16.

99. "SIU Class Construct Huge Dome," *Alton Evening Telegraph* (Feb. 19, 1962), p. 9.

100. "Pretty Mt. V. Girl in a Whatsit," *Mt. Vernon Register-News* (Nov. 19, 1959), p. 17.

101. Croessman, John H., "40-year-old fairground domes gets makeover," *DuQuoin Evening Call* (DuQuoin, IL) (Sept. 27, 2013).

102. Mitchell, B., *Carbondale: A Pictorial History* (St. Louis: G. Bradley, 1991), p. 185.

103. Building permit application, City of Carbondale (Carbondale, IL) (Sept. 16, 1975).

104. Baldwin, *op. cit.*, p. 128.

105. "Space Exhibition Will Be On View Through Summer," *Clarion-Ledger* (Jackson, MS) (June 25, 1959), p. 23.

106. "SJS Brings Dr. Buckminster Fuller In As First Scholar-in-Residence," *Santa Cruz Sentinel* (Jan. 30, 1966), p. 33.

107. Hatch, *op. cit.*, p. 205.

108. "SIU Prof Designs Covered Diamond for Tokyo Giants," *Journal Gazette* (Mattoon, IL) (Jan. 31, 1961), p. 5.

109. "World Citizens Colony Planned," *Lubbock Avalanche-Journal* (Dec. 17, 1970), p. 12.

110. Marx, B., "Dome over Park Ave., not only ignored proposal," *Southern Illinoisan* (Feb. 13, 1983), p. 19.

111. "Fuller building good idea," *Southern Illinoisan* (Dec. 5, 1967), p. 4.

112. "Cone Sweet Home," *Southern Illinoisan* (April 18, 1965), p. 5.

113. Markey, Frank Jay, "Pull Up a Chair," *Pampa Daily News* (Pampa, TX) (Jan. 4, 1968), p. 14.

114. "Glass City Proposed for East Side," *St. Louis Post-Dispatch* (Feb. 26, 1971), p. 4.

115. Hatch, *op. cit.*, p. 242.

116. *Ibid.*

117. *Ibid.*
118. Hatch, *op. cit.*, p. 243.
119. Gelman, B., "Let's Play...", *op. cit.*, p. 3.
120. Niceley, M. A., "World game," *Southern Illinoisan* (Mar. 29, 1970), p. 8.
121. Gelman, B., "Let's Play...," *op. cit.*, p. 3.
122. Mitchell correspondence.
123. Gelman, B., "Antiques Fit Just Fine in Fullers' Dome Home," *Southern Illinoisan* (Dec. 4, 1960), p. 5.
124. Applewhite, p. 13
125. White and Borgognoni report.
126. Gelman, "Antiques Fit...," *op. cit.*, p. 5.
127. *Ibid.*
128. Mitchell correspondence.
129. Hatch, *op. cit.*, p. 220.
130. *Ibid.*
131. Hatch, *op. cit.*, p. 219.
132. Gelman, "Antiques Fit...," *op. cit.*, p. 5.
133. *Ibid.*
134. *Ibid.*
135. "AREA, NOTICE," *Southern Illinoisan* (Dec. 26, 1967), p. 15.
136. Hatch, *op. cit.*, p. 220.
137. Gelman, "Antiques Fit...," *op. cit.*, p. 5.
138. Interview with Shirley Maine, Aug. 11, 2015.
139. "The Dymaxion American," *op. cit.*, p. 50.
140. Gelman, B., "'Bucky' Fuller leaves behind another dome accomplishment," *Southern Illinoisan* (Apr. 24, 1988), p. 35.
141. Interview with Thomas Zung, Sept. 15, 2015.
142. *Ibid.*
143. Sharkey interview.
144. Interview with Michael Ben-Eli, Sept. 9, 2015.
145. *Ibid.*
146. Ben Gelman Collection (c. 1940-1999), Southern Illinois University Special Collections, Morris Library, Carbondale, Illinois.
147. Ritter, G., "FBI Buckminster Fuller file released," *Carbondale Times* (May 22, 2015).
148. Interview with Jamie Snyder, Sept. 11, 2015.
149. *Ibid.*
150. *Ibid.*
151. *Ibid.*
152. Interview with Alexandra Snyder, Oct. 19, 2015.
153. Jamie Snyder interview.
154. *Ibid.*
155. Alexandra Snyder interview.
156. Jamie Snyder interview.
157. *Ibid.*
158. *Ibid.*
159. *Ibid.*
160. *Ibid.*
161. Baldwin, *op. cit.*, p. 125.
162. Heckman interview.
163. *Ibid.*
164. "Businessman dies, car runs through fence," *Southern Illinoisan* (Jan. 22, 1968), p. 3.
165. Interview with Robert Childs, August 28, 2015.
166. *Ibid.*
167. Interview with Ann Dillard, July 1, 2015.
168. Maine interview.
169. Interview with Patrick Lillard, Aug. 26, 2015.
170. Interview with Robbie Stokes, Oct. 27, 2015.
171. Interview with Gary Wallace, Oct. 29, 2015.
172. Interview with Lisa Poppen, Sept. 15, 2015.
173. Interview with Dylan (Poppen) Caraker, Sept. 26, 2015.
174. *Ibid.*
175. *Ibid.*
176. Interview with Toni Roan, Dec. 14, 2015.
177. Interview with Barbara Roan, Dec. 14, 2015.
178. *Ibid.*
179. Toni Roan interview.
180. Barbara Roan interview.
181. *Ibid.*
182. "The Dymaxion American," *op. cit.*, p. 46+.
183. "The Fuller Legacy," *Southern Alumni* (Mar. 2008), p. 15.
184. *Reader's Guide to Periodic Literature #24* (New York: H.W. Wilson Co., 1965), p. 796; *Reader's Guide to Periodic Literature #29* (New York: H.W. Wilson Co., 1969), p. 473.
185. "'Fuller World' TV Programs to be Aired," *Southern Illinoisan* (Oct. 6, 1964), p. 3; "Merv Griffin," *Tucson Daily Citizen* (Nov. 16, 1965), p. 13; "Merv Griffin Show," *Independent* (Long Beach, CA) (Mar. 12, 1968), p. 36.
186. Credits for the TV movie *Earth II*, can be found at imdb.com/title/tt0067039/?ref_=nm_flmg_act_106
187. Mitchell correspondence.
188. Rickerl, S., "A transformative effort," *The Southern* (Apr. 20, 2010).
189. The website of the Buckminster Fuller Institute, BFI.org, has a chronological list of all Fuller-obtained U.S. patents: bfi.org/about-fuller/bibliography/patents
190. A list of many of the awards and honors Fuller received during this lifetime can be found at the following: bfi.org/about-fuller/biography
191. "Fuller to receive area appreciation award," *Southern Illinoisan* (Sept. 26, 1967), p. 2.
192. A list of many of the awards and honors Fuller received during this lifetime can be found at the following: bfi.org/about-fuller/biography
193. Perk interview.
194. Zung interview.

Chapter 2

1. Koplowitz, H. B., *Carbondale After Dark* (Hollywood, CA: Dome Publications, 1982), pp. 18-20.
2. Hatch, A., *Buckminster Fuller at Home in the Universe* (New York: Dell), 1974, p. 252.
3. Chu, H.-Y., *New Views on Buckminster Fuller* (Palo Alto, CA: Stanford University Press), 2009, p. 19.
4. Smith, W., "Odd Man Out," *Chicago Tribune* (May 22, 1995), p. C1.
5. Chu, *op. cit.*, p. 18.
6. *Ibid.*, p 19.

7. Copies of correspondence between Fuller and representatives of SIUC's Morris Library, dating between May 25, 1960 and Aug. 27, 1972, can be found at Morris Library Special Collections, Southern Illinois University at Carbondale, Carbondale, IL.

8. "Fuller plans office on SIU-Edwardsville Campus," *Southern Illinoisan* (Sept. 24, 1971), p. 1.

9. McCue, G., "R. Buckminster Fuller Shifting to Edwardsville Campus," *St. Louis Post-Dispatch* (Sept. 24, 1971), p. 23.

10. "Fuller plans office...," p. 1.

11. Gelman, B., "SIU's budget proposal revised," *Southern Illinoisan* (Feb. 18, 1972), p. 1.

12. Butler, D., "Fuller leaves SIU-Carbondale," *Southern Illinoisan* (July 2, 1972), p. 2.

13. Chu, *op. cit.*, p. 19.

14. Interview with Thomas Zung, Sept. 15, 2015.

15. Interview with Shirley Sharkey, Apr. 17, 2016.

16. *Ibid.*

17. McCue, G., "The Symposium as a Force in Design," *St. Louis Post-Dispatch* (June 18, 1961), p. 76.

18. "Religious Center Set for 1970," *Edwardsville Intelligencer* (Edwardsville, IL) (Aug. 15, 1969), p. 3.

19. Information about SIUE's Fuller dome is available at their website: siue.edu/religion/about_dome.shtml

20. *Ibid.*

21. Hatch, *op. cit.*, p. 253.

22. *Ibid.*, p. 254.

23. "SPACIOUS HOUSING," *Southern Illinoisan* (Oct. 28, 1972), p. 25.

24. "SPACIOUS HOUSING," *Southern Illinoisan* (Nov. 7, 1972), p. 26.

25. From late 2015 until its publication, Mike "Mickey" Mitchell, the second owner of Carbondale's "Bucky Dome," exchanged hundreds of e-mails with the authors of this book. This information is gleamed from that correspondence.

26. *Ibid.*

27. "The Dymaxion American," *Time* (Jan. 10, 1964), p. 50.

28. Smith, *op. cit.*, p. C1.

29. Gowan, A., *Shared Vision: The Second American Bauhaus* (Cambridge, MA: Merrimack, 2012).

30. "The Dymaxion American," *op. cit.*, p. 50.

31. Mitchell correspondence.

32. *Ibid.*

33. *Ibid.*

34. *Ibid.*

35. *Ibid.*

36. *Ibid.*

37. Interview with John Christy, Sept. 10, 2015.

38. *Ibid.*

39. "The Dome is Home," *Southern Illinoisan* (June 23, 1974), p. 16.

40. *Ibid.*

41. Mitchell correspondence.

42. *Ibid.*

43. *Ibid.*

44. *Ibid.*

45. *Ibid.*

46. *Ibid.*

47. *Ibid.*

48. *Ibid.*

49. *Ibid.*

50. Interview with Dr. Bruce Hector, Sept. 15, 2015.

51. *Ibid.*

52. *Ibid*

53. Interview with Jeff Doherty, Dec. 1, 2015.

54. *Ibid.*

55. Mitchell correspondence.

56. Interview with Michelle Bach, Dec. 5, 2015.

57. *Ibid.*

58. *Ibid.*

59. *Ibid.*

60. *Ibid.*

61. *Ibid.*

62. *Ibid.*

63. *Ibid.*

64. "BACH'S GALLERY," *Southern Illinoisan* (Aug. 30, 1976), p. 13.

65. Bach interview.

66. "Houses for Rent," *Southern Illinoisan* (May 17, 1978), p. 28.

67. "Houses for Rent," *Southern Illinoisan* (Nov. 20, 1978), p. 19.

68. Block, A., "Want to live in a geodesic dome? Bucky Fuller's is for rent," *Southern Illinoisan* (Nov. 29, 1978), p. 67.

69. Mitchell correspondence.

70. "Linda Burzynski of 407 S. Forest in Carbondale...," *Southern Illinoisan* (Mar. 13, 1979), p. 18.

71. "Linda Ann Burzynski" (obituary), *Southern Illinoisan* (Dec. 15, 2006).

72. Interview with Ed Cook, Sept. 7, 2015.

73. *Ibid*

74. *Ibid.*

75. *Ibid.*

76. *Ibid.*

77. *Ibid.*

78. Interview with H. B. Koplowitz, July 13, 2015.

79. *Ibid.*

80. *Ibid.*

81. Koplowitz, H. B., "Living in Bucky's Dome," *Southern Alumni* (Mar. 2008), p. 16.

82. *Ibid.*

83. *Ibid.*

84. *Ibid.*

85. *Ibid.*

86. Koplowitz interview.

87. Koplowitz, "Living in Bucky's Dome," *op. cit.*, p. 16.

88. *Ibid.*

89. Interview with Gary Marx, Jan. 8, 2016.

90. *Ibid.*

91. *Ibid.*

92. Interview with Deborah Browne, Mar. 3, 2016.

93. *Ibid.*

94. *Ibid.*

95. *Ibid.*

96. Interview with Lisa Poppen, Sept. 15, 2015.

97. Interview with Dylan (Poppen) Caraker, Sept. 26, 2015.

98. Interview with Bill Perk, Nov. 16, 2015.

99. Hatch, *op. cit.*, p. 256.

100. Sharkey interview.

101. Butler, D., "Bucky related ideas at SIU luncheon," *Southern Illinoisan* (May 6, 1973, p. 3.

102. deFiebre, H., "Overflow crowd cheers 'Bucky' at SIU-C," *Southern Illinoisan* (Feb. 24, 1976), p. 3; "Fuller to be Earth Week speaker at SIU," *Southern Illinoisan* (Apr. 23, 1980), p. 12.

103. "Anne Fuller Dies at 87, Buckminster's Widow," *Los Angeles Times* (July 4, 1983), p. 44.

104. Findagrave.com: findagrave.com/memorial/1403/r.-buckminster-fuller

105. Interview with Donald S. Wham, Oct. 5, 2015.

106. *Ibid.*

107. *Ibid.*

108. Interview with Paul Hansel, Nov. 22, 2015.

109. *Ibid.*

110. *Ibid.*

111. *Ibid.*

112. *Ibid.*

113. *Ibid.*

114. *Ibid.*

115. Interview with William Clay, Nov. 22, 2016.

116. *Ibid.*

117. *Ibid.*

118. *Ibid.*

119. *Ibid.*

120. *Ibid.*

121. *Ibid.*

122. *Ibid.*

123. *Ibid.*

124. *Ibid.*

125. Carbondale Code Department: Non-Traffic Complaint and Notice to Appear (Sept. 15, 1986).

126. Clay interview.

127. *Ibid.*

128. *Ibid.*

129. *Ibid.*

130. *Ibid.*

131. *Ibid.*

132. *Ibid.*

133. "Homes for Sale," *Southern Illinoisan* (May 13, 1987), p. 19.

134. "BUCKMINSTER FULLER'S GEODESIC DOME...," *Southern Illinoisan* (Mar. 13, 1988), p. 29.

135. Interview with Linda Buchholz, Nov. 11, 2015.

136. *Ibid.*

137. *Ibid.*

138. *Ibid.*

139. *Ibid.*

140. *Ibid.*

141. *Ibid.*

142. *Ibid.*

143. *Ibid.*

144. *Ibid.*

145. Mattmiller, B., "'Buckyphiles' expected to pack into Fuller reception," *Southern Illinoisan* (Mar. 19, 1990), p. 3.

146. Buchholz interview.

147. Mitchell correspondence.

148. *Ibid.*

149. *Ibid.*

150. "Real Estate for Sale," *Southern Illinoisan* (Dec. 26, 1994), p. 21.

151. Interview with Cheryl Ginsberg, July 22, 2015.

152. *Ibid.*

153. Mitchell correspondence.

154. *Ibid.*

155. *Ibid.*

156. Perk interview.

157. *Ibid.*

158. *Ibid.*

159. *Ibid.*

160. *Ibid.*

161. *Ibid.*

162. *Ibid.*

163. *Ibid.*

164. *Ibid.*

165. *Ibid.*

166. *Ibid.*

167. *Ibid.*

168. Interview with Linda Gladson, July 7, 2015.

169. *Ibid.*

170. Letter from Michael S. Mitchell to the City of Carbondale (June 12, 1996).

171. Mitchell correspondence.

172. *Ibid.*

173. "Proclamation," *Southern Illinoisan* (July 5, 1995), p. 44.

174. Perk interview.

175. *Ibid.*

176. *Ibid.*

177. *Ibid.*

178. *Ibid.*

179. Interview with Barbara Zieba, Feb. 12, 2016.

180. Bode, G., "Buyers question actual cost of geodesic dome," *Daily Egyptian* (Aug. 8, 1996).

181. Perk interview.

182. Mitchell correspondence.

183. Carbondale Building & Neighborhood Services Division-Housing Inspection Report, Carbondale, IL (Feb. 17, 1995).

184. Interview with John Yow, Dec. 14, 2015.

185. Smith, *op. cit.*, p. C1.

186. Interview with Kim DePaoli, Dec. 21, 2015.

187. *Ibid.*

188. Online posting by Kim DePaoli, Apr. 4, 1995: cjfearnley.com/geodesic-logs/1995/LOG9504.txt

189. DePaoli interview.

190. *Ibid.*

191. *Ibid.*

192. *Ibid.*

193. *Ibid.*

194. Interview with Cornelius Crane, Aug. 8, 2015.

195. *Ibid.*
196. *Ibid.*
197. *Ibid.*
198. Interview with Colleen Doyle, Oct. 20, 2017.
199. *Ibid.*
200. *Ibid.*
201. Binder, S., "Bucky dome might be new home," *Southern Illinoisan* (June 18, 1996), p. 1.
202. Doyle interview.
203. *Ibid.*
204. *Ibid.*
205. Interview with Martin Moran, Apr. 2, 2016.
206. *Ibid.*
207. *Ibid.*
208. *Ibid.*
209. *Ibid.*
210. *Ibid.*
211. *Ibid.*
212. *Ibid.*
213. Interview with Matt Staulcup, Dec. 12, 2015.
214. *Ibid.*
215. Interview with Troy C. Smith, Sept. 5, 2015.
216. Interview with Alexis Holle, Sept. 9, 2015.
217. *Ibid.*
218. *Ibid.*
219. *Ibid.*
220. *Ibid.*
221. *Ibid.*
222. Smith interview.
223. *Ibid.*
224. *Ibid.*
225. *Ibid.*
226. *Ibid.*
227. Holle interview.
228. Smith interview.
229. *Ibid.*
230. Holle interview.
231. Interview with William Denton, Oct. 10, 2017.
232. *Ibid.*
233. *Ibid.*
234. *Ibid.*
235. *Ibid.*
236. *Ibid.*
237. Interview with Kevin Connor, Jan. 9, 2016.
238. *Ibid.*
239. *Ibid.*
240. *Ibid.*
241. *Ibid.*
242. *Ibid.*
243. Mitchell correspondence.
244. Crane interview.
245. Interview with Brent Ritzel, June 8, 2016.
246. Interview with Thad Heckman, July 5, 2015.
247. Mitchell correspondence.
248. *Ibid.*
249. Perk interview.
250. *Ibid.*
251. *Ibid.*
252. *Ibid.*
253. *Ibid.*
254. Mitchell correspondence.
255. *Ibid.*
256. Connor interview.
257. *Ibid.*
258. Mitchell correspondence.

Chapter 3

1. Interview with Bill Perk, Aug. 23, 2015.
2. Interview with Cornelius Crane, Aug. 8, 2015.
3. Bode, G., "The R. Buckminster Fuller Dome, the former residence of the well-known late Bucky Fuller,...." *Daily Egyptian* (Apr. 20, 1995).
4. *Ibid.*
5. Bode, G., "Bucky Dome named to endangered historic sites list," *Daily Egyptian* (Apr. 1, 2004).
6. Interview with Thad Heckman, Aug. 1, 2017.
7. *Ibid.*
8. *Ibid.*
9. *Ibid.*
10. Pearlman, B. B., "Fallingwater at risk of falling," *Sun Journal* (Baltimore, MD) (Nov. 8, 2000).
11. Dwyer, D., "When they built the Hancock Tower—and it started falling apart," *Boston Globe* (July 15, 2015).
12. Heckman interview.
13. From late 2015 until its publication, Mike "Mickey" Mitchell, the second owner of Carbondale's "Bucky Dome," exchanged hundreds of e-mails with the authors. This information is gleamed from that correspondence.
14. Heckman interview.
15. Lau, W., "The Restoration of Buckminster Fuller's Dome Home Kicks off Saturday," *Architect Magazine* (Apr. 18, 2014).
16. *Ibid.*
17. Heckman interview.
18. Bode, "Bucky Dome named...," *op. cit.*
19. Perk interview.
20. *Ibid.*
21. Interview with Justin R. Edgren, Oct. 27, 2015.
22. *Ibid.*
23. *Ibid.*
24. *Ibid.*
25. *Ibid.*
26. Binder, S., "Bucky's Legacy," *The Southern* (Oct. 27, 2001).
27. *Ibid.*
28. Perk interview.
29. Interview with Blair Wolfram, Aug. 3, 2015.
30. *Ibid.*
31. *Ibid.*
32. *Ibid.*
33. Binder, *op. cit.*
34. *Ibid.*
35. *Ibid.*
36. Heckman interview.
37. *Ibid.*

38. Edgren interview.
39. Perk interview.
40. Interview with Joe Clinton, Sept. 14, 2015.
41. *Ibid.*
42. *Ibid.*
43. Binder, S., *op. cit.*
44. Crane interview.
45. *Ibid.*
46. *Ibid.*
47. *Ibid.*
48. *Ibid.*
49. *Ibid.*
50. *Ibid.*
51. Perk interview.
52. Crane interview.
53. Heckman interview.
54. Krutsinger, L. A., "…And Bucky Fuller: Former SIU Visionary's Ideas on View at the Mall," *The Southern* (Feb. 14, 2002).
55. Interview with Joni Reeves, Aug. 26, 2015.
56. *Ibid.*
57. *Ibid.*
58. *Ibid.*
59. *Ibid.*
60. *Ibid.*
61. *Ibid.*
62. *Ibid.*
63. Perk interview.
64. Reeves interview.
65. Interview with Judy Ashby, July 14, 2015.
66. *Ibid.*
67. *Ibid.*
68. Interview with Larry Weatherford, Oct. 10, 2015.
69. *Ibid.*
70. *Ibid.*
71. *Ibid.*
72. *Ibid.*
73. Interview with Larry Busch, Aug. 28, 2015.
74. Heckman interview.
75. Perk interview.
76. Interview with Lisa Keorkenmeier, July 23, 2015.
77. *Ibid.*
78. Bode, G., "Buckminster Fuller dome becomes Carbondale historic landmark," *Daily Egyptian* (Nov. 19, 2003).
79. Kampwerth, A., "Dome Group Fights for National Historic Recognition," *The Southern* (Feb. 11, 2004).
80. Kampwerth, A., "Preservation and Learning," *The Southern Illinoisan* (Dec. 2, 2003), p. 29.
81. Kampwerth, A., "Dome Group Fights…," *op. cit.*
82. Bode, G., "Bucky Dome named…," *op. cit.*
83. *Ibid.*
84. Perk interview.
85. White and Borgognoni Architects, Buckminster Fuller Dome Home Evaluation Report (Carbondale, IL), 2006.
86. *Ibid.*
87. *Ibid.*
88. *Ibid.*
89. *Ibid.*
90. *Ibid*
91. Mitchell correspondence.
92. Wolfram interview.
93. Heckman interview.
94. *Ibid.*
95. *Ibid.*
96. White and Borgognoni report.
97. *Ibid.*
98. Heckman interview.
99. *Ibid.*
100. *Ibid.*
101. *Ibid.*
102. *Ibid.*
103. *Ibid.*
104. Wolfram interview.
105. Bode, G., "Bucky Dome named…," *op. cit.*
106. Crane interview.
107. Bradley, J., "Auction to raise cash for 'Bucky' dome," *The Southern* (Nov. 20, 2003).
108. Reeves interview.
109. Cates, K., "'Bucky dome' finds new home on historic places register," *The Southern* (Mar. 19, 2006).
110. Bode, G., "Buckminster Fuller Dome Home named historic landmark district," *Daily Egyptian* (Oct. 10, 2003).
111. White and Borgognoni report.
112. *Ibid.*
113. Erickson, K., "Architects name top 150 great places in Illinois," *The Southern* (Mar. 25, 2007).
114. Swanson, S., "In a small world, his dome is still big," *Chicago Tribune* (July 5, 2008).
115. Testa, Adam, "Life and legacy of 'Bucky' to debut in Chicago," *The Southern* (Apr. 28, 2009), p. 20.
116. *Ibid.*
117. "Local artists band together to restore Fuller home with compilation CD," *The Southern* (Dec. 29, 2005).
118. *Ibid.*
119. *Roam Home To A Dome*, RBF Dome NFP (2005).
120. Interview with Kathy Livingston, Aug. 5, 2015.
121. *Ibid.*
122. *Ibid.*
123. *Ibid.*
124. *Ibid.*
125. Interview with Judy Ashby, Feb. 15, 2016 (second interview).
126. *Ibid.*
127. Bode, G., "Embrace the 'Dark,'" *Daily Egyptian* (Apr. 19, 2007).
128. "Businesses band together for Bucky's Dome," *The Southern* (Aug. 14, 2008).
129. Bode, G., "Orlandi Vineyard to host second Bucky benefit," *Daily Egyptian* (June 18. 2009).
130. Reeves interview.
131. Thomas, B., "Bucky Dome fans push for landmark status," *The Southern* (July 24, 2009).
132. Fitzgerald, S., "Preserving Bucky's Dome," *The Southern Illinoisan* (May 16, 2008), p. 12.
133. *Ibid.*
134. "Another view of the dome" (Letters to the Editor),

The Southern Illinoisan (May 20, 2008), p. 4.

135. Fitzgerald, *op. cit.*, p. 12.

136. Krajelis, B., "Arbor District treats people to sweets, sales," *The Southern* (Aug. 26, 2007).

137. Information about the National Historic Landmark designation and process can be found at the website: NPS.gov/nhl.

138. *Ibid.*

139. *Ibid.*

140. *Ibid.*

141. Interview with Janet Donoghue, Jan. 14, 2016.

142. *Ibid.*

143. *Ibid.*

144. *Ibid.*

145. *Ibid.*

146. Interview with Brent Ritzel, Feb. 21, 2016.

147. *Ibid.*

148. *Ibid.*

149. *Ibid.*

150. *Ibid.*

151. *Ibid.*

152. *Ibid.*

153. *Ibid.*

154. *Ibid.*

155. *Ibid.*

156. *Ibid.*

157. *Ibid.*

158. *Ibid.*

159. *Ibid.*

160. "Fuller Dome will get major grant for restoration," *The Southern* (Feb. 3, 2011).

161. Ritzel interview.

162. *Ibid.*

163. Weatherford interview.

164. "Dr. Linda Marie Hostalek" (obituary), *The Southern* (June 27, 2018).

165. Heckman interview.

166. *Ibid*

167. *Ibid.*

168. *Ibid.*

169. *Ibid.*

170. *Ibid.*

171. *Ibid.*

172. *Ibid.*

173. *Ibid.*

174. *Ibid.*

175. *Ibid.*

176. *Ibid.*

177. *Ibid.*

178. Realworldfx.com

179. Interview with Todd Ulrich, Sept. 10, 2015.

180. *Ibid.*

181. *Ibid.*

182. Interview with Michael Ulrich, Sept. 21, 2015.

183. Todd Ulrich interview.

184. *Ibid.*

185. "Can't Stand in the Corner" by Stace England (2006/ Stace England).

186. "Recovering the Dome 2008," youtube.com/ watch?v=YKHeDfTLY-c

187. Todd Ulrich interview.

188. *Ibid.*

189. Malkovich, B., "Thousands still without power," *Southern Illinoisan* (May 13, 2009), p. 4.

190. *Ibid.*

191. *Ibid.*

192. Todd Ulrich interview.

193. Heckman interview.

194. *Ibid.*

195. Bode, G., "Fuller Dome not condemned, seeks funding for renovation," *Daily Egyptian* (Apr. 22, 2009).

196. *Ibid.*

197. *Ibid.*

198. Rickerl, S., "A transformative effort," *The Southern* (Apr. 20, 2010).

199. Ritzel, B., "Spotlight: Fuller Dome Home Turns 50," *The Southern* (Jan. 26, 2010).

200. *Ibid.*

201. *Ibid.*

202. Ritzel interview.

203. Heckman interview.

204. Interview with Brian Gorecki, Aug. 3, 2015.

205. *Ibid.*

206. *Ibid.*

207. Interview with Ed Cook, July 22, 2015.

208. *Ibid.*

209. *Ibid.*

210. Interview with Shannon McDonald, Aug. 26, 2015.

211. *Ibid.*

212. *Ibid.*

213. *Ibid.*

214. Interview with Ben Lowder, Sept. 2, 2015.

215. *Ibid.*

216. *Ibid.*

217. *Ibid.*

218. "Fuller Dome will get major...," *op. cit.*

219. Donoghue interview.

220. Heckman interview.

221. *Ibid.*

222. Interview with Josh Rucinski, Sept. 18, 2015.

223. *Ibid.*

224. *Ibid.*

225. *Ibid.*

226. Donoghue interview.

227. *Ibid.*

228. *Ibid.*

229. Rush, L., "Hoping for a big 2011," *The Southern* (Jan. 10, 2011).

230. Rodriguez, Codell, "Rebuild the Dome," *The Southern* (Apr. 10, 2011).

231. Information about the Holland Prize can be found at: nps.gov/hdp/competitions/holland.htm

232. *Ibid.*

233. Heckman interview.

234. Rodriguez, "Rebuild the....," *op. cit.*

235. Interview with Terry Hickey, May 30, 2016.

236. Interview with Peter Bahn, May 25, 2016.

237. Interview with Jessica Allee, Nov. 20, 2016.

238. *Ibid.*

239. *Ibid.*

240. *Ibid.*
241. Interview with Stephen Schauwecker, Mar. 29, 2016.
242. *Ibid.*
243. *Ibid.*
244. Fitzgerald, S., "Buck Up for the Bucky Dome," *Southern Illinoisan* (June 20, 2012), p. 13.
245. Norris, D. W., "Council bats tourism money around," *The Southern* (Feb. 22, 2012).
246. Ritzel interview.
247. Heckman interview.
248. Ritzel interview.
249. *Ibid.*
250. "Two men killed in plane crash identified," KHQA Newsdesk (Hannibal, MO) (Aug. 30, 2012).
251. *Ibid.*
252. "John Johnson" (obituary), *The Southern* (Sept. 5, 2012).
253. *Ibid.*
254. "Bucky Dome effort receives $1,500 grant," *The Southern* (Jan. 30, 2013).
255. Heckman interview.
256. "Bucky Dome effort…," *op. cit.*
257. "Dome work set for spring," *The Southern* (Jan. 2, 2013), p. 14.
258. "SIU exhibit focuses on Bucky," *The Southern* (Apr. 1, 2013).
259. Interview with Jon Davey, Feb. 2, 2016.
260. *Ibid.*
261. Norris, D. W., "Bucky Fuller 'Dome Days' start today," *The Southern* (Apr. 18, 2013).
262. Interview with Mel Goot, Aug. 10, 2015.
263. Interview with Stephen Moore, Jan. 12, 2017.
264. *Ibid.*
265. *Ibid.*
266. "Invitation to Bidders," *Southern Illinoisan* (July 10, 2013), p. 31.
267. Heckman interview.
268. *Ibid.*
269. Davis, T., "Fuller Dome Home to be fully restored," *Daily Egyptian*, Apr. 7, 2014.
270. *Ibid.*
271. Wolfram interview.
272. *Ibid.*
273. Heckman interview.
274. Stout, R., "Restoring the Beauty & Integrity of Fuller's Geodesic Dome," Carlisle Syntec Systems (Aug. 31, 2015), p. 1.
275. Wolfram interview.
276. *Ibid.*
277. Davey interview.
278. Interview with Peter Szczesniweski, Mar. 19, 2016.
279. *Ibid.*
280. *Ibid.*
281. *Ibid.*
282. Fitzgerald, S., "Bucky dome restoration project lifts off—finally," *The Southern* (Apr. 20, 2014).
283. "Friend of Buckminster Fuller strives to rehab the inventor's dome," *St. Louis Post-Dispatch* (Apr. 29, 2002), p. 12.
284. Cates, *op. cit.*

285. "R. Buckminster Fuller group adds board members, continues raising funds," *The Southern* (Oct. 19, 2011).
286. Donoghue interview.
287. Interview Jay Needham, Sept. 2, 2015.
288. *Ibid.*
289. Davey interview.
290. Fitzgerald, S., "Fuller dome restoration nears halfway point," *The Southern* (July 12, 2014).
291. Wolfram interview.
292. Heckman interview.
293. Wolfram interview.
294. *Ibid.*
295. Heckman interview.
296. Interview with Chuck Grammer, Aug. 18, 2015
297. *Ibid.*
298. Interview with Tom Guelcher, Sept. 4, 2015.
299. *Ibid.*
300. *Ibid.*
301. Interview with Bill Quigley, Sept. 16, 2015.
302. Grammer interview.
303. *Ibid.*
304. *Ibid.*
305. *Ibid.*
306. Interview with Braden Larson, Nov. 11, 2015.
307. *Ibid.*
308. *Ibid.*
309. *Ibid.*
310. *Ibid.*
311. Mathis, C., "Fuller's work, legacy are focus of Morris Library gala," *The Southern* (Sept. 29, 2014).
312. Perk interview.
313. Davey interview.
314. Heckman interview.
315. "Architectural firm shares its Top 10," *The Southern* (Jan. 1, 2014).
316. "Fuller discussion to be Oct. 19," *The Southern* (Oct. 18, 2014).
317. Interview with Daniel Presley, June 26, 2017.
318. *Ibid.*
319. *Ibid.*
320. *Ibid.*
321. Interview with Lindy Loyd, Jan. 11, 2017.
322. *Ibid.*
323. *Ibid.*
324. *Ibid.*
325. *Ibid.*
326. Duncan, D., "'Bucky Dome' receives Historic Preservation Award," *The Southern* (May 20, 2015).
327. Motchan, B., "Preservation Efforts Recognized at Landmarks Illinois Driehaus Awards Presentation," *Chicago Architecture* (Oct. 21, 2015).
328. Heckman interview.
329. *Ibid.*
330. *Ibid.*
331. Lane Museum: lanemotormuseum.org/collection/cars/item/dymaxion-replica-1933
332. *Ibid.*
333. Kozak, G., "We drive Buckminster Fuller's terrifying Dymaxion car (so you don't have to)," *Autoweek*

(May 27, 2015).

334. Hetzler, B., "Blast from the Past," *The Southern* (Oct. 23, 2015).

335. Ashby interview (second interview).

336. *Ibid.*

337. *Ibid.*

338. Interview with Thad Heckman, Dec. 15, 2017 (second interview).

339. *Ibid.*

340. *Ibid.*

341. *Ibid.*

342. *Ibid.*

343. Interview with Rick Cantwell, June 8, 2015.

344. *Ibid.*

345. *Ibid.*

346. Heckman interview (second interview).

347. *Ibid.*

348. *Ibid.*

349. *Ibid.*

350. *Ibid.*

351. *Ibid.*

352. *Ibid.*

353. *Ibid.*

354. *Ibid.*

355. *Ibid.*

356. *Ibid.*

357. *Ibid.*

358. *Ibid.*

359. *Ibid.*

360. Meeting minutes: RBF Dome NFP (Oct. 11, 2016).

361. *Ibid.*

362. *Ibid.*

363. *Ibid.*

364. Interview with Ed Cook, July 15, 2018 (second interview).

365. Meeting minutes: RBF Dome NFP (Feb. 28, 2017).

366. *Ibid.*

367. *Ibid.*

368. *Ibid.*

369. *Ibid.*

370. Esch, K. J., "Buckminster Fuller birthday party celebrates inventor's life and legacy," *The Southern* (July 11, 2017).

371. "CARBONDALE: Buckminster Fuller Dome," *Southern Illinoisan* (Nov. 30, 2017).

372. Meeting minutes: RBF Dome NFP (Aug. 23, 2016).

373. Chow, D., "Spotlight on Carbondale: Illinois Town Sits at Solar Eclipse 'Crossroads," LiveScience.com (Aug. 20, 2017) (livescience.com/60179-carbondale-illinois-solar-eclipse-crossroads.html)

374. *Ibid.*

375. *Ibid.*

376. Meeting minutes: RBF Dome NFP (Apr. 4, 2017).

377. *Ibid.*

378. Meeting minutes: RBF Dome NFP (Sept. 9, 2017).

379. Meeting minutes: RBF Dome NFP (Oct. 11, 2016).

380. Meeting minutes: RBF Dome NFP (June 13, 2018).

381. Interview with Stephen Schauwecker, July 1, 2018 (second interview).

382. *Ibid.*

383. *Ibid.*

384. *Ibid.*

385. *Ibid.*

386. *Ibid.*

387. *Ibid.*

388. "Notice to Proceed," RBF Dome NFP, June 25, 2018.

389. Interview with Eric Grosshenrich, Aug. 30, 2018.

390. *Ibid.*

391. *Ibid.*

392. Eric Grosshenrich's Dome restoration videos can be found at the following link: youtube.com/playlist?list=PLgZlIEGb3KSS_B0ppreH6Icktcoch2LBT

393. Meeting minutes: RBF Dome NFP (July 10, 2018).

394. *Ibid.*

395. *Ibid.*

396. *Ibid.*

397. *Ibid.*

398. Interview with Thad Heckman, Dec. 26, 2018 (third interview).

399. *Ibid.*

400. Heckman interview (third interview).

401. Interview with Judy Ashby, Aug. 16, 2018 (third interview).

402. Cook interview (second interview).

403. Heckman interview (third interview).

404. Grosshenrich interview.

405. Ashby interview (third interview).

406. YouTube video, "Dome Restoration: Day 39 – 41": youtube.com/watch?v= N8uCWMXzLnM&index= 38&list=PLgZlIEGb3KSS_B0ppreH6Icktcoch2LBT

407. Meeting minutes: RBF Dome NFP (Sept. 25, 2018).

408. *Ibid.*

409. Heckman interview (third interview).

410. YouTube video, "Dome Restoration: Day 31": youtube.com/watch?v=3VcCaM8_IgQ&index=32&list=PLgZlIEGb3KSS_B0ppreH6Icktcoch2LBT

411. *Ibid.*

412. YouTube video, "Dome Restoration: Day 36 – 38": youtube.com/watch?v= 9RWM_67gGZs&list=PLgZlIEGb3KSS_B0ppreH6Icktcoch2LBT&index=37

413. Cook interview (second interview).

414. Heckman interview (third interview).

415. "Farnsworth House, Wrigley Field among those voted for best building in Illinois," *Kendall County Now* (Apr. 27, 2018).

416. Meeting minutes: RBF Dome NFP (July 10, 2018).

417. "Sturgis Award presented to Davey, Thompson," *SIU News* (Aug. 21, 2018).

418. "Bucky's Daughter & Granddaughter to visit the Dome," Center for Spirituality and Sustainability (SIUE) (Oct. 26, 2018). fullerdome.org/blog/2018/10/26/buckys-daughter-amp-granddaughter-to-visit-the-dome?fbclid=IwAR3GcvR8SOApA27WBmsQknCfjgglRBbSgsphY2IHk8cRA-yNcSqcirN0K2U

419. Interview with Anita Presley, Jan. 8, 2019.

420. *Ibid.*

421. Interview with Christian Baumann, Jan. 9, 2019.
422. *Ibid.*
423. Interview with Thad Heckman, Aug., 2019 (fourth interview).
424. *Ibid.*
425. *Ibid.*
426. *Ibid.*
427. Lowder, Ben, e-mail communication (Jan. 17, 2019).
428. *Ibid.*
429. Heckman interview (fourth interview).
430. *Ibid.*
431. *Ibid.*
432. *Ibid.*
433. Meeting minutes: RBF Dome NFP (June 11, 2019).
434. *Ibid.*
435. *Ibid.*

BIBLIOGRAPHY

"Anne Fuller Dies at 87, Buckminster's Widow," *Los Angeles Times* (July 4, 1983), p. 44.

"Another view of the dome" (Letters to the Editor), *The Southern Illinoisan* (May 20, 2008), p. 4.

Applewhite, E. J., *Cosmic Fishing: An Account of Writing Synergetics with Buckminster Fuller* (New York: Macmillan, 1977), p. 12.

"Architectural firm shares its Top 10," *The Southern* (Jan. 1, 2014).

"AREA, NOTICE," *Southern Illinoisan* (Dec. 26, 1967), p. 15.

"Art Educators Meet at SIU," *Southern Illinoisan* (Nov. 1, 1964), p. 3.

"BACH'S GALLERY," *Southern Illinoisan* (Aug. 30, 1976), p. 13.

Baldwin, J., *Buckyworks* (New York: John Wiley, 1996), p. 40.

Ben Gelman Collection (*c.* 1940–1999), Southern Illinois University Special Collections, Morris Library, Carbondale, Illinois.

BFI.org

Binder, S., "Bucky dome might be new home," *Southern Illinoisan* (June 18, 1996), p. 1; "Bucky's Legacy," *The Southern* (Oct. 27, 2001).

Blades, J., "Russo's no fool," *Chicago Tribune* (Feb. 19, 1995), p. 171.

Block, A., "Want to live in a geodesic dome? Bucky Fuller's is for rent," *Southern Illinoisan* (Nov. 29, 1978), p. 67.

Bode, G., "The R. Buckminster Fuller Dome, the former residence of the well-known late Bucky Fuller,...." *Daily Egyptian* (Apr. 20, 1995); "Buyers question actual cost of geodesic dome," *Daily Egyptian* (Aug. 8, 1996); "Buckminster Fuller Dome Home named historic landmark district," *Daily Egyptian* (Oct. 10, 2003); "Buckminster Fuller dome becomes Carbondale historic landmark," *Daily Egyptian* (Nov. 19, 2003); "Bucky Dome named to endangered historic sites list," *Daily Egyptian* (April 1, 2004); "Embrace the 'Dark,'" *Daily Egyptian* (Apr. 19, 2007); "Fuller Dome not condemned, seeks funding for renovation," *Daily Egyptian* (April 22, 2009); "Orlandi Vineyard to host second Bucky benefit," *Daily Egyptian* (June 18, 2009).

Bradley, J., "Auction to raise cash for 'Bucky' dome," *The Southern* (Nov. 20, 2003).

"Buckminster Fuller Will Speak to NUEA," *Edwardsville Intelligencer* (Oct. 25, 1967), p. 9.

"BUCKMINSTER FULLER'S GEODESIC DOME...," *Southern Illinoisan* (Mar. 13, 1988), p. 29.

"Bucky Dome effort receives $1,500 grant," *The Southern* (Jan. 30, 2013).

"'Bucky' Fuller, Design Revolutionary, Still Super-Charged with Ideas," *Southern Illinoisan* (October 26, 1959), p. 16.

"'Bucky' to picnic with area children," *Southern Illinoisan* (May 26, 1971), p. 2.

"Bucky's Daughter & Granddaughter to visit the Dome," Center for Spirituality and Sustainability (SIUE) (Oct. 26, 2018). fullerdome.org/blog/2018/10/26/buckys-daughter-amp-granddaughter-to-visit-the-dome?fbclid=IwAR3GcvR8SOApA27WBmsQknCfjgglRBbSgsphY2IHk8cRA-yNcSqcirN0K2U

Building permit application, City of Carbondale (Carbondale, IL) (Sept. 16, 1975).

"Businesses band together for Bucky's Dome," *The Southern* (Aug. 14, 2008).

"Businessman dies, car runs through fence," *Southern Illinoisan* (Jan. 22, 1968), p. 3.

Butler, D., "Fuller leaves SIU-Carbondale," *Southern Illinoisan* (July 2, 1972), p. 2; "Bucky related ideas at SIU luncheon," *Southern Illinoisan* (May 6, 1973, p. 3.

"Cambridge City Area First with Geodesic Dome Dealership," *National Road Traveler* (Cambridge City, IN) (Sept. 24, 959), p. 1.

"Can't Stand in the Corner" by Stace England (2006/Stace England).

"CARBONDALE: Buckminster Fuller Dome," *Southern Illinoisan* (Nov. 30, 2017).

Carbondale Building & Neighborhood Services Division-Housing Inspection Report, Carbondale, IL (Feb. 17, 1995).

Carbondale Code Department: Non-Traffic Complaint and Notice to Appear (Sept. 15, 1986).

Cates, K., "'Bucky dome' finds new home on historic places register," *The Southern* (Mar. 19, 2006).

Chow, D., "Spotlight on Carbondale: Illinois Town Sits at Solar Eclipse 'Crossroads," LiveScience.com (Aug. 20, 2017) (livescience.com/60179-carbondale-illinois-solar-eclipse-crossroads.html)

Chu, H.-Y., *New Views on Buckminster Fuller* (Palo Alto, CA: Stanford University Press), 2009, p. 19.

Croessman, J. H., "40-year-old fairground domes gets makeover," *DuQuoin Evening Call* (DuQuoin, IL) (Sept. 27, 2013).

"Cone Sweet Home," *Southern Illinoisan* (April 18, 1965), p. 5.

Davis, T., "Fuller Dome Home to be fully restored," *Daily Egyptian*, Apr. 7, 2014.

deFiebre, Henry, "Overflow crowd cheers 'Bucky' at SIU-C," *Southern Illinoisan* (Feb. 24, 1976), p. 3; "Fuller to be Earth Week speaker at SIU," *Southern Illinoisan* (Apr. 23, 1980), p. 12.

"Dome Home For Fuller," *Southern Illinoisan* (Oct. 21, 1959), p. 1.

"Dome Home is Erected in Seven Hours," *Southern Illinoisan* (Apr. 20, 1960), p. 3.

"Dome Home Number Two Is Going Up," *Southern Illinoisan* (May 10, 1960), p. 2.

"Dome work set for spring," *The Southern* (Jan. 2, 2013), p. 14.

"Domes for Designers," *Southern Illinoisan* (Dec. 6, 1960), p. 16.

"Dr. Linda Marie Hostalek" (obituary), *The Southern* (June 27, 2018).

Duncan, D., "'Bucky Dome' receives Historic Preservation Award," *The Southern* (May 20, 2015).

Dwyer, D., "When they built the Hancock Tower—and it started falling apart," *Boston Globe* (July 15, 2015).

E-mail communication to the authors from Maureen Berkowitz, Jackson County Assessment Office, June 23, 2015.

Erickson, K., "Architects name top 150 great places in Illinois," *The Southern* (Mar. 25, 2007).

Esch, K. J., "Buckminster Fuller birthday party celebrates inventor's life and legacy," *The Southern* (July 11, 2017).

"Farnsworth House, Wrigley Field among those voted for best building in Illinois," *Kendall County Now* (Apr. 27, 2018).

findagrave.com/memorial/1403/r.-buckminster-fuller

Fitzgerald, S., "Preserving Bucky's Dome," *The Southern Illinoisan* (May 16, 2008), p. 12; "Buck Up for the Bucky Dome," *Southern Illinoisan* (June 20, 2012), p. 13; "Bucky dome restoration project lifts off—finally," *The Southern* (Apr. 20, 2014); "Fuller dome restoration nears halfway point," *The Southern* (July 12, 2014).

"Friend of Buckminster Fuller strives to rehab the inventor's dome," *St. Louis Post-Dispatch* (Apr. 29, 2002), p. 12.

"Fuller building good idea," *Southern Illinoisan* (Dec. 5, 1967), p. 4.

"Fuller discussion to be Oct. 19," *The Southern* (Oct. 18, 2014).

"Fuller Dome will get major grant for restoration," *The Southern* (Feb. 3, 2011).

"Fuller plans office on SIU-Edwardsville Campus," *Southern Illinoisan* (Sept. 24, 1971), p. 1.

"Fuller slated PTA speaker," *Southern Illinoisan* (Sept. 7, 1967), p. 7.

"Fuller to receive area appreciation award," *Southern Illinoisan* (Sept. 26, 1967), p. 2.

"Fuller to Talk Monday," *Southern Illinoisan* (Feb. 7, 1965), p. 2.

"Fuller to Talk on World Design," *Southern Illinoisan* (May 12, 1964), p. 3.

"Fuller Will Talk Jan. 11," *Southern Illinoisan* (Jan. 3, 1963), p. 2.

"'Fuller World' TV Programs to be Aired," *Southern Illinoisan* (Oct. 6, 1964), p. 3.

Gelman, B., "Antiques Fit Just Fine in Fullers' Dome Home," *Southern Illinoisan* (Dec. 4, 1960), p. 5; "Let's Play 'Make the World Work' Fuller Urges," *Southern Illinoisan* (March 10, 1965), p. 3; "Tired? Just think about Bucky's schedule," *Southern Illinoisan* (Apr. 27, 1967), p. 3; "Fuller sees great progress toward 'world man," *Southern Illinoisan* (Sept. 24, 1967), p. 36; "SIU's budget proposal revised," *Southern Illinoisan* (Feb. 18, 1972), p. 1; "'Bucky' Fuller leaves behind another dome accomplishment," *Southern Illinoisan* (Apr. 24, 1988), p. 35.

"Glass City Proposed for East Side," *St. Louis Post-Dispatch* (Feb. 26, 1971), p. 4.

Gowan, A., *Shared Vision: The Second American Bauhaus* (Cambridge, MA: Merrimack, 2012).

Hatch, A., *Buckminster Fuller at Home in the Universe* (New York: Dell, 1974), pp. 218 and 252.

"'Haystack House' Draws Wide Interest," *Southern Illinoisan* (May 27, 1960), p. 3.

"'Haystack House' Termed Coming Thing in House," *Hartford Courant* (May 15, 1960), p. 10D.

Hetzler, B., "Blast from the Past," *The Southern* (Oct. 23, 2015).

"Homes for Sale," *Southern Illinoisan* (May 13, 1987), p. 19.

"Houses for Rent," *Southern Illinoisan* (May 17, 1978), p. 28.

"Houses for Rent," *Southern Illinoisan* (Nov. 20, 1978), p. 19.

imdb.com/title/tt0067039/?ref_=nm_flmg_act_106

"Invitation to Bidders," *Southern Illinoisan* (July 10, 2013), p. 31.

"John Johnson" (obituary), *The Southern* (Sept. 5, 2012).

Kampwerth, A., "Dome Group Fights for National Historic Recognition," *The Southern* (Feb. 11, 2004); "Preservation and Learning," *The Southern Illinoisan* (Dec. 2, 2003), p. 29.

Koplowitz, H. B., "Living in Bucky's Dome," *Southern Alumni* (March 2008), p. 16; *Carbondale After Dark* (Hollywood, CA: Dome Publications, 1982), pp. 18-20.

Kozak, G., "We drive Buckminster Fuller's terrifying Dymaxion car (so you don't have to)," *Autoweek* (May 27, 2015).

Krajelis, B., "Arbor District treats people to sweets, sales," *The Southern* (Aug. 26, 2007).

Krutsinger, L. A., "...And Bucky Fuller: Former SIU Visionary's Ideas on View at the Mall," *The Southern* (Feb. 14, 2002).

Lane Museum: lanemotormuseum.org/collection/cars/item/dymaxion-replica-1933

Lau, W., "The Restoration of Buckminster Fuller's Dome Home Kicks off Saturday," *Architect Magazine* (Apr. 18, 2014).

Letter from Michael S. Mitchell to the City of Carbondale (June 12, 1996).

"Linda Ann Burzynski" (obituary), *Southern Illinoisan* (Dec. 15, 2006).

"Linda Burzynski of 407 S. Forest in Carbondale...," *Southern Illinoisan* (Mar. 13, 1979), p. 18.

"Local artists band together to restore Fuller home with compilation CD," *The Southern* (Dec. 29, 2005).

Lowder, B., e-mail communication (Jan. 17, 2019).

Malkovich, B., "Thousands still without power," *Southern Illinoisan* (May 13, 2009), p. 4.

Markey, F. J., "Pull Up a Chair," *Pampa Daily News* (Pampa, TX) (Jan. 4, 1968), p. 14.

Marx, B., "Dome over Park Ave., not only ignored proposal," *Southern Illinoisan* (Feb. 13, 1983), p. 19.

Mathis, C., "Fuller's work, legacy are focus of Morris Library gala," *The Southern* (Sept. 29, 2014).

Mattmiller, B., "'Buckyphiles' expected to pack into Fuller reception," *Southern Illinoisan* (Mar. 19, 1990), p. 3.

McCue, G., "R. Buckminster Fuller Shifting to Edwardsville Campus," *St. Louis Post-Dispatch* (Sept. 24, 1971), p. 23.

McCue, G., "The Symposium as a Force in Design," *St. Louis Post-Dispatch* (June 18, 1961), p. 76.

Meeting minutes: RBF Dome NFP, Aug. 23, 2016; Oct. 11, 2016; Feb. 28, 2017; Apr. 4, 2017; Sept. 9, 2017; Jun. 13, 2018; Jul. 10, 2018; Sept. 25, 2018; Jun. 11, 2019.

"Merv Griffin," *Tucson Daily Citizen* (Nov. 16, 1965), p. 13.

"Merv Griffin Show," *Independent* (Long Beach, CA) (Mar. 12, 1968), p. 36.

Mitchell, B., *Carbondale: A Pictorial History* (St. Louis: G. Bradley, 1991), p. 185; *Delyte Morris of SIU* (Carbondale, IL: SIU Press, 1988), p. 94.

Motchan, B., "Preservation Efforts Recognized at Landmarks Illinois Driehaus Awards Presentation," *Chicago Architecture* (Oct. 21, 2015).

Niceley, M. A., "World game," *Southern Illinoisan* (Mar. 29, 1970), p. 8.

Norris, D. W., "Bucky Fuller 'Dome Days' start today," *The Southern* (Apr. 18, 2013); "Council bats tourism money around," *The Southern* (Feb. 22, 2012).

"Notice to Proceed," RBF Dome NFP, June 25, 2018.

NPS.gov/nhl.

O'Dell, C., *Virginia Marmaduke: A Journey in Print from Carbondale to Chicago* (Chicago: Arcadia, 2001), p. 13.

Olive, R., "Dome Provides Light Structures," *Eugene Guard* (Eugene, OR) (Jan. 18, 1959), p. 26.

Online posting by Kim DePaoli, Apr. 4, 1995: cjfearnley.com/geodesic-logs/1995/LOG9504.txt

Pearlman, B. B., "Fallingwater at risk of falling," *Sun Journal* (Baltimore, MD) (Nov. 8, 2000).

Pease Domes brochure, Pease Manufacturing Company (Hamilton, OH), 1959.

Potter, R., *Buckminster Fuller* (Englewood Cliffs, NJ: Silver Burdett, 1990), p. 104.

Powell, L., "An engineering whodunit at Black Mountain College," *North Carolina Miscellany* (Oct. 13, 2017).

"Pretty Mt. V. Girl in a Whatsit," *Mt. Vernon Register-News* (Nov. 19, 1959), p. 17.

"Proclamation," *Southern Illinoisan* (July 5, 1995), p. 44.

"R. Buckminster Fuller group adds board members, continues raising funds," *The Southern* (Oct. 19, 2011).

Reader's Guide to Periodic Literature #24 (New York: H.W. Wilson Co., 1965), p. 796; *Reader's Guide to Periodic Literature #29* (New York: H.W. Wilson Co., 1969), p. 473.

Realworldfx.com

"Real Estate for Sale," *Southern Illinoisan* (Dec. 26, 1994), p. 21.

"Recovering the Dome 2008," youtube.com/watch?v=YKHeDfTLY-c

"Religious Center Set for 1970," *Edwardsville Intelligencer* (Edwardsville, IL) (Aug. 15, 1969), p. 3.

Rickerl, S., "A transformative effort," *The Southern* (Apr. 20, 2010).

Ritter, G., "FBI Buckminster Fuller file released," *Carbondale Times* (May 22, 2015).

Ritzel, B., "Spotlight: Fuller Dome Home Turns 50," *The Southern* (Jan. 26, 2010.)

Roam Home To A Dome, RBF Dome NFP (2005).

Rodriguez, C., "Rebuild the Dome," *The Southern* (Apr. 10, 2011).

Rush, L., "Always on call," *Southern Illinoisan* (April 22, 2010); "Hoping for a big 2011," *The Southern* (Jan. 10, 2011).

"Scenic Design Contest Planned," *Southern Illinoisan* (Jan. 7, 1962), p. 19.

Schnyder, M., "National history museum showcases little-known Wichita artifact." *Wichita Eagle* (Dec. 19, 2016).

siu.edu/about-siu/history.php

"SIU Appoints Fuller," *Southern Illinoisan* (Sept. 20, 1959), p. 3.

"SIU Business Writers Meet Starts Today," *Southern Illinoisan* (Apr. 9, 1965), p. 2.

"SIU Class Construct Huge Dome," *Alton Evening Telegraph* (Feb. 19, 1962), p. 9.

"SIU exhibit focuses on Bucky," *The Southern* (Apr. 1, 2013).

"SIU Prof Designs Covered Diamond for Tokyo Giants," *Journal Gazette* (Mattoon, IL) (Jan. 31, 1961), p. 5.

"SJS Brings Dr. Buckminster Fuller In As First Scholar-in-Residence," *Santa Cruz Sentinel* (Jan. 30, 1966), p. 33.

Smith, W., "Odd Man Out," *Chicago Tribune* (May 22, 1995), p. C1.

"Space Exhibition Will Be On View Through Summer," *Clarion-Ledger* (Jackson, MS) (June 25, 1959), p. 23.

"SPACIOUS HOUSING," *Southern Illinoisan* (Nov. 7, 1972), p. 26.

"SPACIOUS HOUSING," *Southern Illinoisan* (Oct. 28, 1972), p. 25.

Stout, R., "Restoring the Beauty & Integrity of Fuller's Geodesic Dome," Carlisle Syntec Systems (Aug. 31, 2015), p. 1.

"Sturgis Award presented to Davey, Thompson," *SIU News* (Aug. 21, 2018).

Swanson, S., "In a small world, his dome is still big," *Chicago Tribune* (July 5, 2008).

Testa, A., "Life and legacy of 'Bucky' to debut in Chicago," *The Southern* (Apr. 28, 2009), p. 20.

"The Dome is Home," *Southern Illinoisan* (June 23, 1974), p. 16.

"The Dymaxion American," *Time* (Jan. 10, 1964), pp. 46 and 50.

"The Fuller Legacy," *Southern Alumni* (Mar. 2008), p. 15.

Thomas, B., "Bucky Dome fans push for landmark status," *The Southern* (July 24, 2009).

"Two men killed in plane crash identified," KHQA Newsdesk (Hannibal, MO) (Aug. 30, 2012).

University Archives, Morris Library, Southern Illinois University at Carbondale: lib.siu.edu/collections/scrc/scrc-collections/university-archives.php

White and Borgognoni Architects, "Buckminster Fuller Dome Home Evaluation Report" (Carbondale, IL), 2006.

"World Citizens Colony Planned," *Lubbock Avalanche-Journal* (Dec. 17, 1970), p. 12.

YouTube video, "Dome Restoration: Day 31": youtube.com/watch?v=3VcCaM8_IgQ&index=32&list=PLgZlIEGb3KSS_B0ppreH6Icktcoch2LBT

YouTube video, "Dome Restoration: Day 36 – 38": youtube.com/watch?v=9RWM_67gGZs&list=PLgZlIEGb3KSS_B0ppreH6Icktcoch2LBT&index=37

YouTube video, "Dome Restoration: Day 39 – 41": youtube.com/watch?v=N8uCWMXzLnM&index=38&list=PLgZlIEGb3KSS_B0ppreH6Icktcoch2LBT

Interviews

Interview with Alexandra Snyder, Oct. 19, 2015.
Interview with Alexis Holle, Sept. 9, 2015.
Interview with Anita Presley, Jan. 8, 2019.
Interview with Ann Dillard, July 1, 2015.
Interview with Barbara Roan, Dec. 14, 2015.
Interview with Barbara Zieba, Feb. 12, 2016.
Interview with Ben Lowder, Sept. 2, 2015.
Interview with Bill Perk, Aug. 23, 2015; Nov. 16, 2015.
Interview with Bill Quigley, Sept. 16, 2015.
Interview with Blair Wolfram, Aug. 3, 2015.
Interview with Braden Larson, Nov. 11, 2015.
Interview with Brent Ritzel, Feb. 21, 2016; June 8, 2016.
Interview with Brian Gorecki, Aug. 3, 2015.
Interview with Cheryl Ginsberg, July 22, 2015.
Interview with Christian Baumann, Jan. 9, 2019.
Interview with Chuck Grammer, Aug. 18, 2015
Interview with Colleen Doyle, Oct. 20, 2017.
Interview with Cornelius Crane, Aug. 8, 2015.
Interview with Daniel Presley, June 26, 2017.
Interview with Deborah Browne, Mar. 3, 2016.
Interview with Donald S. Wham, Oct. 5, 2015.
Interview with Dr. Bruce Hector, Sept. 15, 2015.
Interview with Dylan (Poppen) Caraker, Sept. 26, 2015.
Interview with Ed Cook, July 22, 2015; Sept. 7, 2015; July 15, 2018.
Interview with Eric Grosshenrich, Aug. 30, 2018.
Interview with Gary Marx, Jan. 8, 2016.
Interview with Gary Wallace, Oct. 29, 2015.
Interview with H. B. Koplowitz, July 13, 2015.
Interview with Harold Cohen, Nov. 20, 2015.
Interview with Herb Meyer, Sept. 18, 2015.
Interview with Ira Parrish, Oct. 1, 2015.
Interview with Jamie Snyder, Sept. 11, 2015.
Interview with Janet Donoghue, Jan. 14, 2016.
Interview with Jay Needham, Sept. 2, 2015.
Interview with Jeff Doherty, Dec. 1, 2015.
Interview with Jessica Allee, Nov. 20, 2016.
Interview with Joe Clinton, Sept. 14, 2015.
Interview with John Christy, Sept. 10, 2015.
Interview with John Yow, Dec. 14, 2015.
Interview with Jon Davey, Feb. 2, 2016.
Interview with Joni Reeves, Aug. 26, 2015.
Interview with Josh Rucinski, Sept. 18, 2015.
Interview with Judy Ashby, July 14, 2015; Feb. 15, 2016; Aug. 16, 2018.
Interview with Justin R. Edgren, Oct. 27, 2015.
Interview with Kathy Livingston, Aug. 5, 2015.
Interview with Kevin Connor, Jan. 9, 2016.
Interview with Kim DePaoli, Dec. 21, 2015.
Interview with Larry Busch, Aug. 28, 2015.
Interview with Larry Weatherford, Oct. 10, 2015.
Interview with Linda Buchholz, Nov. 11, 2015.
Interview with Linda Gladson, July 7, 2015.
Interview with Lindy Loyd, Jan. 11, 2017.
Interview with Lisa Keorkenmeier, July 23, 2015.
Interview with Lisa Poppen, Sept. 15, 2015.
Interview with Martin Moran, Apr. 2, 2016.

Interview with Matt Staulcup, Dec. 12, 2015.
Interview with Mel Goot, Aug. 10, 2015.
Interview with Michael Ben-Eli, Sept. 9, 2015.
Interview with Michael Ulrich, Sept. 21, 2015.
Interview with Michelle Bach, Dec. 5, 2015.
Interview with Patrick Lillard, Aug. 26, 2015.
Interview with Paul Hansel, Nov. 22, 2015.
Interview with Peter Bahn, May 25, 2016.
Interview with Peter Szczesniweski, Mar. 19, 2016.
Interview with Rick Cantwell, June 8, 2015.
Interview with Robbie Stokes, Oct. 27, 2015.
Interview with Robert Childs, Aug. 28, 2015.
Interview with Shannon McDonald, Aug. 26, 2015.
Interview with Shirley Maine, Aug. 11, 2015.
Interview with Shirley Sharkey, Apr. 17, 2016.
Interview with Stephen Moore, Jan. 12, 2017.
Interview with Stephen Schauwecker, Mar. 29, 2016; Jul. 1, 2018.
Interview with Terry Hickey, May 30, 2016.
Interview with Thad Heckman, Jul. 5, 2015; Aug. 1 and Dec. 15, 2017; Dec. 26, 2018; Aug. 2019.
Interview with Thomas Zung, Sept. 15, 2015.
Interview with Todd Ulrich, Sept. 10, 2015.
Interview with Tom Guelcher, Sept. 4, 2015.
Interview with Toni Roan, Dec. 14, 2015.
Interview with Troy C. Smith, Sept. 5, 2015.
Interview with William Clay, Nov. 22, 2016.
Interview with William Denton, Oct. 10, 2017.